AF659059

Crop Diversification and Soil Health Management for Sustainable Development

The Editors

Dr. Sanjay-Swami is Professor in Soil Science and Agricultural Chemistry at School of Natural Resource Management, College of Post Graduate Studies in Agricultural Sciences, Barapani under Central Agricultural University, Imphal. He holds a Ph.D. in Soil Science from CCS Haryana Agricultural University, Hisar, India. He is involved in teaching UG and PG courses. He is an active researcher in the field of Conservation and Management of Natural Resources, especially in the Indian Himalayan Region for more than 15 years. He handled 5 externally funded projects and contributed over 60 research papers and technical articles in various journals/periodicals of international repute. He has also to his credit ten standard books and more than 45 book chapters. He has guided/guiding 5 Ph.D. and 11 M.Sc. scholars, attended more than 20 international conferences, presented research papers and won many prestigious awards *viz.* SCSI Leadership Award, Young Scientist Award, Distinguished Scientist Award, Outstanding Achievement Award, Bharat Shiksha Ratan Award, *etc.* He has wide international exposure and visited many countries like The Netherlands, China, Nepal, Bangladesh, Thailand, Mongolia, *etc.* for various academic purposes. His Excellency, The President of India and the Visitor of Manipur University nominated him as a Member of the Fifth Court of Manipur University for a term of three years. He is actively serving as the Chairman of Meghalaya State Chapter of the Soil Conservation Society of India and Member of the International Advisory Council of The Grassroots Institute, Montreal, Canada.

Dr. Sanjay Arora, Ph.D. in Soils Science from Punjab Agricultural University, Ludhiana, Punjab, India. He has more than 22 years experience in the field of soil and water management and is presently Principal Scientist (Soil Science) at Regional Research Station of Central Soil Salinity Research Institute at Lucknow, Uttar Pradesh, India. He has made significant contributions in sustainable management of soil and water resources. He is active member of various academic and professional societies/associations. He is author of 142 peer-reviewed research papers and technical articles that has been published in various national and international journals. He has authored eleven books and contributed 42 book chapters. He has guided 13 M.Sc. students for research related to soil health management. He has experience of handling 7 externally funded research/ consultancy projects and he is recipient of Young Scientist Award, SCSI Leadership Award, Groundwater Augmentation Award, 2009 (MOWR), Dr. R.N. Dwivedi Medal, National Fellow, SCSI and Fellow, SES, J.S. Bali Award, 2016. He is Chief Editor, Journal of Soil and Water Conservation as well as Editor of Journal of the Indian Society of Soil Science, International Journal of Forest, Soil and Erosion, Iran; Journal of Indian Society of Soil Salinity and Water Quality and Associate Editor, International Journal of Microbiology Research. He is Councillor and Indian representative of World Association of Soil and Water Conservation (WASWAC).

Dr. Vikas Sharma is Professor and Head, Division of Soil Science and Agricultural Chemistry at the Sher-e-Kashmir University of Agricultural Sciences and Technology of Jammu, India. He obtained his doctorate from the Punjab Agricultural University, Ludhiana, India. His research interests include soil and water conservation, land degradation, GIS and soil health. Most of his work pertains to soils in Indian Himalayas. He has worked on improving the moisture retention in the moisture deficit *Kandi* areas of Jammu through both *in situ* and *ex situ* management of water, soil erosion risk mitigation, carbon pools and nutrient dynamics in the erosion prone foothill Himalayas. He is also involved in the teaching and advising students' research. He has over 50 Research articles and 15 book chapters to his credit.

Dr. Anshuman Kohli is an Associate Professor at the Department of Soil Science and Agricultural Chemistry, Bihar Agricultural University, Sabour, Bhagalpur, Bihar, India. He holds a Ph.D. in Soil Science from the Punjab Agricultural University, Ludhiana, India (2002). He is actively involved in teaching undergraduate and post graduate students. He has been teaching the post graduate courses on soil physics and soil erosion and conservation since last 10 years. His research and extension activities revolve around participatory research on rainfed cropping systems, rainwater harvesting, indigenous technical knowledge, soil physical quality, water management and precision nutrient management under rainfed environments. His current areas of interest include elucidating the effects of management practices on water and nutrient balance and carbon dynamics in rainfed environments and the drivers of livelihood for resource poor farmers in constrained ecologies. He has authored over 20 peer reviewed research articles and organized several scientific and extension events.

Mr. Shubham Singh is doctoral scholar in the discipline of Soil Science and Agricultural Chemistry at School of Natural Resource Management, College of Post Graduate Studies in Agricultural Sciences, Central Agricultural University, Barapani, Meghalaya. He is working on nutrient management aspect in acidic soils and published six research papers and eight book chapters. He has also been conferred the *Best Master Thesis Award 2019* by Agricultural Technology Development Society during 3rd International Conference at Tribhuvan University, Kathmandu, Nepal for his excellent research on temporal phosphorus availability with *Azolla* incorporation, the corresponding mechanisms and changes in soil acidity indices in acid Inceptisol of Meghalaya under supervision of Prof. Sanjay Swami.

Crop Diversification and Soil Health Management for Sustainable Development

– *Editors* –

Sanjay-Swami

Sanjay Arora

Vikas Sharma

Anshuman Kohli

Shubham Singh

2021

GENE-TECH BOOKS

New Delhi - 110 002

© 2021 EDITORS
ISBN 978-81-89729-56-1

All rights reserved. Including the right to translate or to reproduce this book or parts thereof except for brief quotations in critical reviews.

The authors are solely responsible for the contents of the bock chapters compiled in this book. The editors or publisher do not take any, responsibility for same in any manner. Errors, if any are purely unintentional and readers are requested to communicate such errors to the editors or publisher to avoid discrepancies in future.

Complied and Published on behalf of
Academy of Natural Resource Conservation and Management (ANRCM), Lucknow, U.P., India
http://www.anrcm.org

Published by: **GENE-TECH BOOKS®**
4762-63/23, Ansari Road, Darya Ganj,
New Delhi - 110 002
Phone: 09899295259, 011-43003222
E-mail: genetechbooks@yahoo.co.in

Typeset at: **Classic Computer Services**
Delhi - 110 035

Printed at: **Replika Press Pvt. Ltd.**

PRINTED IN INDIA

Preface

The definition of soil health under various cropping systems has evolved with the development of agriculture. In the past, researchers and farmers were mostly concerned about soil quality and crop production. Since the 1990s, the concept of soil health assessment has focused on specific soil properties and the soil's ability to maintain a range of ecological functions in its appropriate ecosystem, supporting long-term sustainable cropping systems. Thus, soil health is defined as the ability of a soil to function and provide ecosystem services, or the soil's fitness to support crop growth without degrading soil or otherwise harming the environment.

Crop diversification is not only critical for optimizing crop production but also important for increasing soil health by balancing soil biodiversity, enhancing soil nutrient use efficiency, and reducing soil-borne pathogen. The overall richness of crop species in agro-ecosystems could be the ultimate driver of soil health; thus, to optimize the benefits that crop diversification can bring to the system, the diversity of plant functional groups may be important for crop diversification management.

In agro-ecosystems, the most common functional crop mixtures consist of a mixture of any of the four main groups, namely C3 grasses (such as cotton), C4 grasses (such as maize), legumes which fix N from atmosphere, and non-leguminous forbs, as plants with different eco-functional types often grow well in community due to their different needs in the temporal and spatial niche and soil nutrient availabilities. Similarly, higher diversity of plant eco-functional groups creates heterogeneity of the favorable niches for different soil functional microbes; therefore, crop diversification management with more plant functional groups could enhance soil health and ecosystem services.

In modern agriculture, growing new crop varieties with improved compatibility of beneficial soil biota could be a powerful way to improve soil health in agro-ecosystems, as plant genotypes can significantly influence soil microbial communities and their functionalities in agro-ecosystems.

This publication is the outcome of the technical papers presented during the International Web Conference on *Resource Management and Biodiversity Conservation to Achieve Sustainable Development Goals* under the patronage of the Academy of Natural Resource Conservation and Management (ANRCM), Lucknow (UP), India during 11th to 12th September, 2020. The compilation of the papers includes various issues of global concern as well as different regions of the country.

We believe that the publication would surely serve as a valuable source book for the researchers, scientists, teachers, academicians, policy planners and students who want to be fully acquainted with the recent developments in crop diversification and soil health management for sustainable development. We express our sincere thanks to all the contributors for their cooperation and sharing their valuable experiences and thoughts in making this book highly informative. The editors have special words of appreciation to one and all who had helped directly or indirectly in publication of this book.

Sanjay-Swami

Sanjay Arora

Vikas Sharma

Anshuman Kohli

Shubham Singh

Contents

Chapter 1

Influence of Lime and Organic Manures on Growth and Yield Performance of Baby Corn (*Zea mays* L.) under Rainfed Condition

Abdul Rahman Munib[1]*, *K. Pathak*[2], *N. Borah*[3], and *Jami Naveen*[4]

[1]*M.Sc. Student*, [4]*Ph.D. Scholar*,
Department of Agronomy, Assam Agricultural University, Jorhat, Assam
[2]*Principal Scientist, Directorate of Research (Agri), Assam Agricultural University, Jorhat, Assam*
[3]*Professor, Department of Soil Science, Assam Agricultural University, Jorhat, Assam*

A field experiment was conducted at the Assam Agricultural University, Jorhat during the summer season of 2019 to study the effect of lime and manures on growth of baby corn under rainfed condition. The treatment consisted of seven different sources of nutrients viz.,Control, Enriched compost(EC) @ 2.5 t/ha, FYM @2.5 t/ha,Lime (50 per cent LR) + FYM @2.5 t/ha + ash (2.5 kg/ha), Lime (50 per cent LR) + EC @2.5 t/ha + ash (2.5 kg/ha), EC @2.5 t/ha + Lime (25kg/ha) + ash (2.5 kg/ha) mixture at sowing and 30 days after sowing (DAS) and FYM @2.5 t/ha + Lime (25kg/ha) + ash (2.5 kg/ha) mixture at sowing and 30 DAS. The experiment was laid out in RBD with four replications and soil of the experimental field was medium in available nitrogen, potassium and low in available P_2O_5 with pH value of 5.2 and O.C. value of 0.62 per cent. G-5414 was used as the test variety of baby corn. Among the treatments, application of enriched compost (2.5 t ha^{-1}) +lime (25 kg ha^{-1}) +ash (2.5 kgha^{-1}) mixture at sowing and 30 DAS has shown the significantly better results *with respect to growth attributes (viz., plant height, number of leaves, dry matter accumulation, leaf area index, leaf area duration) and finally the highest cob yield(42.88 q ha^{-1}) followed by FYM (2.5 t ha^{-1}) +lime (25 kg ha^{-1}) +ash (2.5 kgha^{-1}) mixture at sowing and 30 DAS (36.63 q ha^{-1}).*

Keywords: *Baby corn, Organic, Lime, Enriched compost, Growth, Dry matter accumulation.*

Maize (*Zea mays* L.) is the third most important cereal crop next to rice and wheat (Murdia *et al.*, 2016). Besides, its use as feed and fodder, maize is of several uses *viz.*, maize flour or meal and corn oil is one of the best cooking oil. It is rich in unsaturated fatty acids. Considering the diversification and value addition of maize as well as the growth of food processing industries, a recent development is of growing maize for the vegetable purpose, which is commonly known as baby corn. Baby corn is the small (6-7 cm long), young, unfertilized corn ear harvested at the stage of silk emergence (*i.e.* within 2 - 3 days of silk emergence). Soil acidity is a serious constraint for crop production in many regions of the world including India. Acid soils around 30 per cent of global arable land, with a growing trend. In the north-eastern region of India, approximately 95 per cent soils are acidic, and nearly 65 per cent soils are suffering from strong acidity with pH less than 5.5 (Thakuria *et al.*, 2016).Crop production in acid soils is hindered substantially by Al and/or Mn toxicity and low levels of available Ca and P. To overcome the problem of acid soil infertility, several agricultural practices have been recommended. Among them, the most common and widely used method is liming. Evidences clearly show that soil application of organic manures either alone or in combination with agricultural lime can directly neutralize the soil acidity.

Today, organic agriculture is receiving increased attention and organic food markets are also expanding at a quite faster rate in many parts of the world including India. This organic market expansion makes it possible for farmers to reap the benefit of a trade with higher price premiums (Ummyiah *et al.*, 2017).In the present scenario, the use of chemical fertilizers may help in achieving maximum yield of baby corn but they pose a serious health hazards and reduces the microbial population in rhizosphere soil, and it is a big threat to sustainability of agriculture.

Development of production technology, particularly appropriate nutrient management strategy for baby corn particularly in acid soils is necessary for realizing higher yield and economic returns (Naveen and Saikia, 2020). Under such circumstances, present investigation was taken up to evaluate the effect of lime and manures in baby corn under rain-fed condition.

Methodology

A field was carried out in the Instructional-cum-Research (ICR) Farm, Assam Agricultural University, Jorhat during summer season of 2019 to evaluate the effect of lime and manures in baby corn under rain-fed condition. The treatment consisted of seven different sources of nutrients *viz.*, T_1: Control, T_2: Enriched compost, EC (2.5 t/ha), T_3: FYM (2.5 t/ha), T_4: Lime (50 per cent LR) + FYM (2.5 t/ha) + ash (2.5 kg/ha), T_5: Lime (50 per cent LR) + EC (2.5 t/ha) + ash (2.5 kg/ha), T_6: EC (2.5 t/ ha) + Lime (25kg/ha) + ash (2.5 kg/ha) mixture at sowing and 30 days after sowing

(DAS) and **T_7:** FYM (2.5 t/ha) + Lime (25kg/ha) + ash (2.5 kg/ha) mixture at sowing and 30 DAS. The experiment was laid out in a randomized block design (RBD) with four replications. The soil of the experimental site was sandy loam in texture, medium in available N (285.36 kgha^{-1}), P_2O_5 (22.85 kgha^{-1}) and K_2O (138.04 kgha^{-1}) and organic carbon (0.62 per cent) with pH value of 5.2. The baby corn variety "G-5414" was used. Sowing of seeds was done manually at a spacing of 45 cm x 20 cm on 20th March 2019and harvesting by picking was started on 30th May 2019 and completed on 17th June 2019. The rainfall received during the crop season was only 680 mm. The weekly average maximum temperature ranged from 24.9 to 32.9°C and minimum temperature 16°C to 26.1°C, respectively.

Results and Discussion

Growth Parameters

The growth of baby corn was measured in terms of plant height, number of leaves plant^{-1}, leaf area index (LAI), leaf area duration (LAD) and dry matter production plant^{-1} at different stages which varied significantly under different nutrient management practices.In all growth stages except 20 DAS, the highest values of plant height and number of leaves plant^{-1} were recorded with application of EC (2.5 t ha^{-1}) + lime (25 kg ha^{-1}) + ash (2.5 kg ha^{-1}) mixture at sowing and 30 DAS (T_6), and differed significantly over other treatments (Table 1.1). This might be due to the fact that enriched compost (EC) had the highest NPK content as compared to other sources used in this study and coupled with lime which increases soil pH, Ca concentration, cation exchange capacity (CEC) and base saturation, simultaneously lowering the Al concentration and increasing P availability (Jafer and Hailu, 2017).

Leaf area index is of paramount importance in all crops, because of optimum leaf area required for a maximum light interception, which results in higher photosynthesis (Boote *et al.*, 1996). The values of leaf area index and leaf area duration were higher with application of EC (2.5 t ha^{-1}) + lime (25 kg ha^{-1}) + ash (2.5 kg ha^{-1}) mixture at sowing and 30 DAS(T_6), which can be attributed to supply of more nitrogen through enriched compost compared to other treatments. With the increased supply of nitrogen, there was a subsequent increase in both LAI and LAD (Kumar· and Singh, 2001). Leaf area index decides the efficiency of canopy photosynthesis. The significantly higher leaf area index due to application of EC (2.5 t ha^{-1}) + lime (25 kg ha^{-1}) + ash (2.5 kg ha^{-1}) mixture at sowing and 30 DAS (T_6) might have resulted in better photosynthesis, thereby contributing to the expression of superior yield attributing characters leading to better yield performance.

Dry matter production is another important character to express the overall growth and metabolic efficiency of plant, which ultimately influences the crop yield. The dry matter production is a result of metabolic activities of the crop plant. Application of EC (2.5 t ha^{-1}) + lime (25 kg ha^{-1}) + ash (2.5 kg ha^{-1}) mixture at sowing and 30 DAS showed significantly superior effects in terms of the total dry matter production of the plants (Figure 1.1). Periodic dry matter production increased with increase in crop age up to maturity. Maximum total dry matter production might be due to more plant height, higher number of leaves plant^{-1}, increased LAI leading to more photosynthetic rate and accumulation of more assimilates. The increased dry

Table 1.1. Effect of Lime and Manures on Plant Height, Number of Leaf per Plant, Leaf Area Index and Leaf Area Duration (LAD) of Baby Corn at 20 Days Interval

Treatment	*Plant Height (cm)*			*No. of Leaves/Plant*			*Leaf Area Index*			*Leaf Area Duration (Days)*		
	20 DAS	*40 DAS*	*60 DAS*	*20 DAS*	*40 DAS*	*60 DAS*	*20 DAS*	*40 DAS*	*60 DAS*	*20 DAS*	*40 DAS*	*60 DAS*
T_1 : Control	26.80	78.30	95.45	4.45	5.95	6.50	2.39	4.81	5.18	23.94	72.04	99.85
T_2 : Enriched compost (2.5 t ha^{-1})	27.55	85.70	106.80	4.60	6.55	7.45	2.41	5.17	5.39	24.10	75.80	105.60
T_3 : FYM (2.5 t ha^{-1})	24.75	83.20	100.65	4.50	6.38	7.23	2.40	5.14	5.26	23.98	75.38	104.03
T_4 : Lime (50 per cent LR) + FYM (2.5 t ha^{-1}) + ash (2.5 kg ha^{-1})	26.30	90.50	106.40	4.60	6.60	7.60	2.41	5.21	5.49	24.13	76.23	106.98
T_5 : Lime (50 per cent LR) + EC (2.5 t ha^{-1}) + ash (2.5 kg ha^{-1})	26.10	92.85	108.70	4.70	6.75	8.10	2.42	5.34	5.61	24.21	77.56	109.48
T_6 : EC (2.5 t ha^{-1}) + lime (25 kg ha^{-1}) + ash (2.5 kg ha^{-1}) mixture atsowing and 30 DAS	25.50	104.05	124.45	4.80	8.70	10.80	2.45	5.63	6.30	24.51	80.76	119.24
T_7 : FYM (2.5 t ha^{-1}) + lime (25 kg ha^{-1}) + ash (2.5 kg ha^{-1}) mixture at sowing and 30 DAS	26.55	97.30	117.15	4.75	7.70	9.30	2.43	5.40	6.21	24.28	78.24	116.03
SEm (±)	1.17	1.47	1.70	0.13	0.14	0.23	0.0038	0.01	0.056	0.038	0.105	0.55

matter production might be due to better utilization of nutrient. This corroborates the findings of Luikham *et al.* (2003).

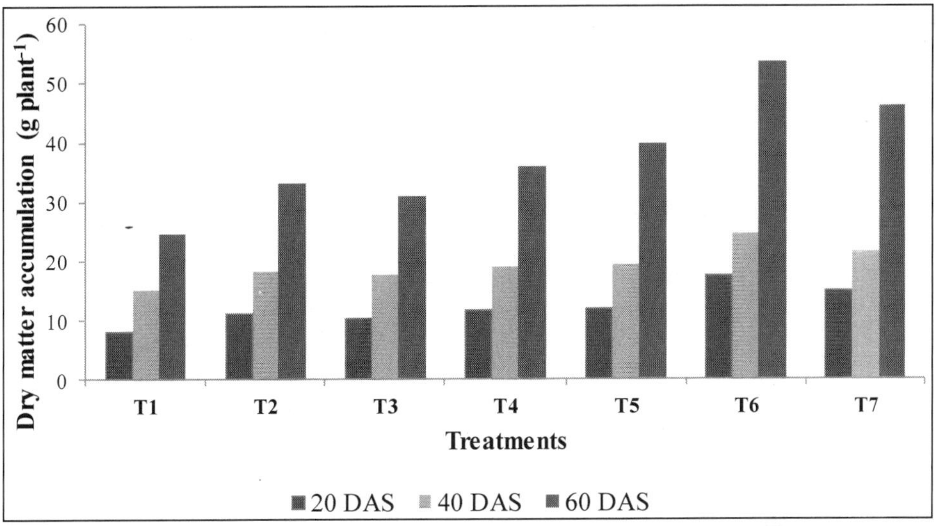

Figure 1.1. Effect of Lime and Manures on Dry Matter Accumulation (g plant^{-1}) of Baby Corn.

Cob Yield and Green Fodder Yield

The highest cob yield was obtained under treatment T_6, *i.e.*, with application of EC (2.5 t ha^{-1}) + lime (25 kg ha^{-1}) + ash (2.5 kg ha^{-1}) mixture at sowing and 30 DAS(T_6) (Figure 1.2). This might be due to the positive combined effect of growth and yield

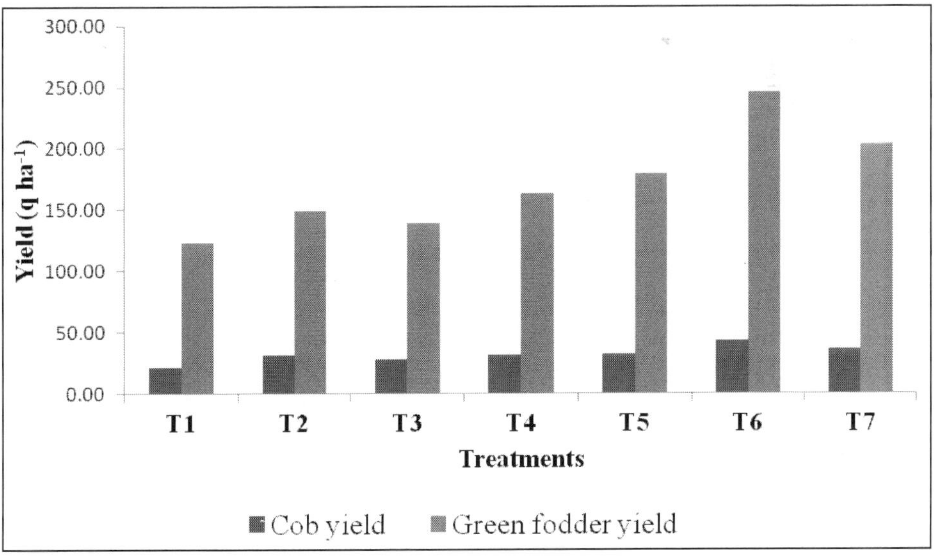

Figure 1.2. Effect of Lime and Manures on Cob Yield and Green Fodder Yield of Baby Corn.

attributing characters. This is in consonance with the findings of Thavaprakash *et al.* (2005), since the yield of the crop is a function of several yield components which are dependent on the complementary interaction between the vegetative and reproductive growth of the crop. Increased nutrient availability *vis-à-vis* enhanced mineralization of organic nitrogen might have increased photosynthetic rate and net assimilation rate, which has resulted in more cob yield. Similar observations were reported by Kar *et al.* (2006) who reported an increase in baby corn yield due to nitrogen application.

Green fodder yield also reflected the superiority of treatment $T_{6,}$ where EC (2.5 t ha^{-1}) + lime (25 kg ha^{-1}) + ash (2.5 kg ha^{-1}) mixture at sowing and 30 DAS was applied (Figure 1.2). Since green fodder yield is a function of plant height, number of leaves and leaf area index and this might have realized in higher green fodder yield in EC (2.5 t ha^{-1}) + lime (25 kg ha^{-1}) + ash (2.5 kg ha^{-1}) mixture at sowing and 30 DAS (T_6). Kar *et al.* (2006) and Dadarwal *et al.* (2009) also reported an increase in fodder yield with a successive increase in nitrogen application.

Conclusion

Thus, the present study showed that application of enriched compost(EC)@2.5 t ha^{-1} with lime (25 kg ha^{-1}) and ash (2.5 kg ha^{-1}) mixture at sowing and 30 days after sowing (DAS) resulted in best performance of baby corn in terms of growth cob yield and green fodder yield indicating the benefit of lime and manure in baby corn.

REFERENCES

Boote, K.J.;Jones, J.W. and Pickering, N.B.(1996). Potential uses and limitations of crop models. *Agronomy J.* **88**(5): 704-716.

Dadarwal, R. S.; Jain, N. K. and Singh, D. (2009).Integrated nutrient management in baby corn (*Zea mays*).*Indian J. Agric. Sci.***79**: 1023-25.

Jafer, D. and Gebresilassie, H. (2017). Application of Lime for Acid Soil Amelioration and Better Soybean Performance in SouthWestern Ethiopia. *Journal of Biology, Agriculture and Healthcare*, 7(5).

Kar, P.P.; Barik, K.C.; Mahapatra, P.K.; Garnayak, L.M.; Rath, B.S.; Bastia, D.K. and Khanda, C.M. (2006). Effect of planting geometry and nitrogen on yield, economics and nitrogen uptake of sweet corn (*Zea mays*). *Indian J. Agron.* **51**(1): 43-45.

Kumar, V. and Singh, K.P.(2001).Enriching vermicompost by nitrogen fixing and phosphate solubilizing bacteria. *Biores. Technol.* **76** (2): 173-175.

Luikham Edwin; Rajan, J. K.; Rajendran, K. and Anal, P. S. M. (2003). Effect of organic and inorganic nitrogen on growth and yield of baby corn (*Zea mays* L.). *Agric. Sci. Digest* **23**: 119-21.

Murdia, L. K., Wadhwani, R., Wadhawan, N., Bajpai, P., Shekhawat, S. (2016). Maize Utilization in India: An Overview. *American Journal of Food and Nutrition*, 4 (6) 169-176.

Naveen, J. and Saikia, M. (2020). Nutrient Management in Organic Baby Corn Production: A Review. *Agril. Rev.*, 41(1): 66-72.

Thavaprakaash, N., Velayudham, K. and Muthukumar, V.B. (2005). Effect of crop geometry, intercropping systems and integrated nutrient management practices on productivity of baby corn (*Zea mays* L.) based intercropping systems. *Res.J.Agric. Biol.Sci.* **1**(4): 295-302.

Thakuria, D., Hazarika, S., and Krishnappa, R. (2016). Soil acidity and management options. *Indian J. Fertil*, **12**, 40-56.

Ummyiah, H. M., Narayan, S., Kumar, P., Nabi, A. and Mudasir Magray, M. A. (2017). Export of organic products: Opportunities and challenges. *J. Pharm. Phytochem.*, 6(6): 1084-1088.

Chapter 2

Bt Cotton in India: An Overview

***Gangishetti Ranjithkumar*[1], *Rajani Bisen*[1] *and Duddukur Rajasekhar*[2]**

[1]Department of Plant Breeding and Genetics, Jawaharlal Nehru Krishi Vishwavidyalaya, Jabalpur, Madhya Pradesh
[2]School of Crop Improvement, College of Post Graduate Studies in Agricultural Sciences, Central Agricultural University (Imphal), Umiam, Meghalaya

Cotton is one of the most important fiber crop in India and it is referred as white gold.After China, India is the largest producer and consumer of cotton. Cotton is a long duration crop and is attacked by large number of insect pests throughout its growth and development and the major problem in cotton cultivation is bollworms i.e American bollworm (Helicoverpa armigera), Pink bollworm (Pectionphora gossypiella) and Spotted bollworm (Earias vitella). Amongst them, Helicoverpa emerged as a key pest all over the country causing as high as 80 per cent losses in cotton and Cotton bolls are highly vulnerable to hidden insects To overcome this problem bt cotton was developed.Bt was first discovered in by Shigetane Ishiwatarias. Earlier Bt has been applied to crops in spray form as aninsecticide, containing a mixture of spores and the associated proteincrystals. Genetic engineering involves in the production of transgenic cotton. Bt- Cotton was produced by using different gene transfer technologies. There are several regulatory authorities are there to control the bio safety issues regarding transgenic cotton. In India Mahyco, in collaboration with Monsanto, introduced Bt cotton technology into India. Bt cotton, the first genetically modified (GM) crop in India, was initially approved in India on March 26th 2002 for commercial cultivation in six states belonging to southern and central cotton cultivation zones of the country.The first approval for commercial cultivation of Bt cotton in India was grantedto three cotton hybrids, MECH-12 Bt, MECH-162 Bt and MECH-184 Bt.

***Keywords**: Bt cotton, Genetic engineering, Cotton bollworms.*

Cotton, generally referred as white gold, it is one of the most important cash crops in India and it plays a key role in the Indian economy since it provides livelihood to more than 60 million people by way of support in agriculture, processing and use of cotton in textiles (Barwale *et al.*, 2004).Since it was first cultivated between 5,000 and 10,000 years ago, cotton production has highly influenced global economic development (Stephens *et al*, 1974). Currently, Cotton is commercially in more than 70 countries including many developing countries around the globe (Zhang BH, Feng R (2000)). As a cash crop Cotton contributes 29.8 per cent of the Indian agricultural gross domestic product, and nearly nine million hectares of land in India is used to produce 14.2 million bales of cotton lint. In cotton, bollworms are major tissue feeders and highly damaging. Three types of bollworms are there *i.e.*, American bollworm (*Helicoverpa armigera*), Pink bollworm (Pectionphora gossypiella) and Spotted bollworm (*Earias vitella*). Amongst them, Helicoverpa emerged as a key pest all over the country causing as high as 80 per cent losses in cotton. To overcome this problem bt cotton was developed.Bt cotton has been already commercialized in six countries the United States (1996), Australia (1997), South Africa (1997), Argentina (1998), Mexico (1996), China (1998), and Indonesia (2000). In India Mahyco (Maharashtra Hybrid Seed Company), in collaboration with Monsanto, introduced Bt cotton technology into India. Bt cotton, the first genetically modified (GM) crop in India, was initially approved in India on March 26th 2002 for commercial cultivation in six states belonging to southern and central cotton cultivation zones of the country.

Introduction to Bt

Bt was first discovered in 1901 by the Japanese biologist Shigetane Ishiwatarias a cause of sotto disease that was killing silkworms and named it Bacillus sotto (Milner 1994). *Bacillus thuringiensis* (Bt) is a common gram positive, spore-forming,soil living bacteria. Upon sporulation, *B. thuringiensis* forms crystals of proteinaceous insecticidal δ-endotoxins (called crystal proteins or Cry proteins), which are encoded by cry genes (Khadi *et al.*, 2002). For over 50 years, Bt has been applied to crops in spray form as an insecticide, containing a mixture of spores and the associated proteincrystals. The development and commercialization of insect-resistant transgenic Bt crops expressing Cry toxins revolutionized the history of agriculture.

Structure of Bt Toxin: Cry Protein Domains

Domain A - Pore formation 6α helices (250 aas)

Domain B -receptor binding domain, β-sheets (200 aas)

Domain C - β sandwich (150 aas)protects the toxin from protease

Cry Proteins Fall into Three Domains According to their Mode of Action

Domain I: causing pore opening in the larvae gut

Domain II: protein binding to gut membranes

Domain III: determining stability and membrane binding

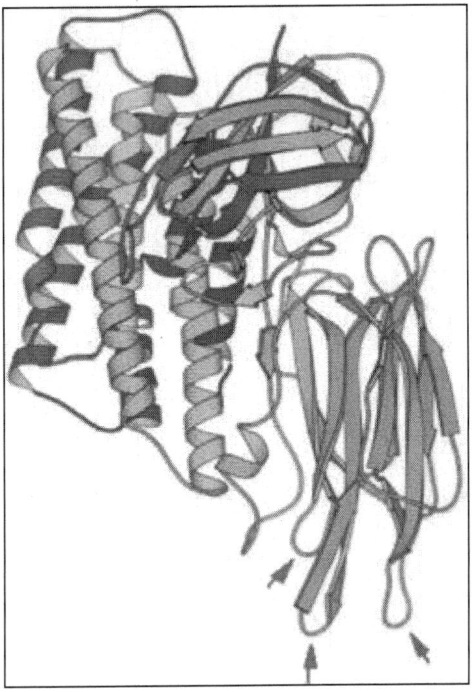

Figure 3.1. A Typical Bt Crystal Protein in Three Dimensional View (*Source*: Kumar *et al.*, 1996).

Table 2.1. Cry Genes and their Functional Orders

Cry Genes	*Toxic to the Insect Order*
Cry1A(a), Cry1A(b), Cry1A(c)	Lepidoptera
Cry1B, Cry1C, Cry1D	Lepidoptera
Cry II	Lepidoptera, Diptera
CryIII	Coleoptera
CryIV	Diptera
CryV	Lepidoptera, Coleoptera

Mode of Action of Bt

Kranthi, K. R. (2012).Well explained in ***Bt cotton-questions and answers*** that Bt produces various protein crystals with spore formation and these are mostly delta- endotoxins that are usually inactive until converted after ingestion by enzymatic action to an active, toxic form in the gut of host larvae. The Cry toxin becomes active by proteoletic enzymes in the alkaline gut juice (pH 8–10).These proteins account for insecticidal activity used against certain members of the moth family, the fly family and the beetle family of insects. Bt also produces other toxins. Microbial Bt preparations have internationally been used on a wide range of plants as bio-pesticides, the first having been produced in France in 1938 and later in the US in 1961 as a registered insecticide spray. The crystal form of the protein toxins are designated by the acronym *Cry*.

The Cry1Ac, Cry2Ab, Cry1C, Cry1F *etc.*, belong to the class 'Bt-delta-endotoxins' and functionally consider as oral toxins. The delta-endotoxins are ingested and the protoxins present in the crystals are proteolytically activated to trypsin-resistant active core δ-endotoxin in the alkaline mid-gut. The active toxin traverses the peritrophic membrane to bind cadherin receptors present on the brush border membrane of the insect midgut. The cadherins process the toxins to form homo-oligomers and bind to specific receptors like alkaline phosphatases and amino peptidases before causing pores in the epithelial membrane, resulting in osmotic lysis of the cells. This results incessation of feeding and finally causing mortality. Amongst the genes that have been deployed in insect resistant transgenic cotton, thus far, Cry1Ac is the most toxic to *H. armigera* and against a wide range of lepidopteran insect pests that include the other two bollworms the spotted bollworm *Earias vittella and* the pink bollworm, *Pectinophora gossypiella*. Bt-cotton incorporated with Cry1Ac is highly toxic to the bollworms and other minor pests such as the cotton semilooper and hairy caterpillar, but not effective on the leaf eating tobacco caterpillar *Spodoptera litura*. Cry2Ab2 (present in Bollgard-II) is moderately toxic to *Helicoverpa armigera*, and *Spodoptera litura*. However the expression levels of Cry2Ab2 are very high (120-350 ppm), as compared to the Cry1Ac (0.5 to 15 ppm) and therefore Bollgard II is highly toxic to *H. armigera* and toxic to *Spodoptera litura*.

Need of Bt Cotton

Cotton is a long duration crop and is attacked by large number of insect pests throughout its growth and development. Before the introduction of Bt cotton, insecticide quantity applied on cotton was the highest, relative to other cultivated crops to control insect pests. Cotton bolls are highly vulnerable to hidden insects such as the American bollworm, pink bollworm and spotted bollworm. Some hidden feeders generally do not come into direct contact with insecticide sprays. Bt Cotton helps to protect the crop from such type of feeders. The first commercial Bt cotton variety was released in USA by M/S. Monsanto (Bollgard), which contains Cry 1Ac gene of *Bacillus thuringiensis*. Bt cotton is commercially grown in several countries like China, Australia, Mexico, South Africa, Argentina, India, Indonesia *etc.* Worldwide the area under Bt cotton keep increasing year by year. Overall, about 12 per cent of the world cotton is now planted with Genetically Modified varieties/ hybrids (GMO) and ICAC has estimated that his may rise to 50 per cent in 5-7 years.

Development of Bt Cotton

The major efforts to harness genetic engineering technology for bollworm resistance in cotton began in the 1990s in India with the import of genetically modified (GM) cotton and initiation of research programmes in national laboratories (Karihaloo and Kumar, 2009).Bt cotton carries the *Cry1Ac* gene derived from the common soil bacterium *Bacillus thuringiensis* var. *kurstaki*, which results in the expression of the *Cry1Ac* protein that confers resistance to the bollworm complex (Barwale *et al.*, 2004).

According to Mayee *et al.*, 2003,For development of any transgenic crop, there are five important steps generally:

(a) Identification of effective gene or genes (Gene of interest)

(b) Gene transfer technology : there are different techniques are there *i.e.*, direct, indirect (Physical, Biological, Chemical)

(c) Regeneration ability from protoplasts, callus or tissues

(d) Gene expression of the product at desired level

(e) Proper integration of genes so that are carried for generations by usual means of reproduction.

Once identification of bollworm inhibiting genes has been achieved, several scientists have step by step solved the problems to achieve perfect transgenics. For development of transgenic plant in cotton, Agrobacterium-mediated gene transfer technique has been essentially used. Now, biolistic gene transfer techniques (One of the direct gene transfer techniques) are available for direct gene transfer to protoplast. The regeneration of cotton plants from callus and somantic embryogenesis have so far been restricted to few 'Coker' genotypes. All cotton genotypes are not amenable to regeneration and that is one big hurdle in gene transfer. There are reports of induction of somantic embryogenesis has also been reported from china and Australia but in India, attempts to repeat it with Indian genotypes have been unsuccessful. Using Cry 1 Ab and Cry 1 Ac genes, transgenic cottons with perfect integration, expression and reproduction was achieved first in USA in 1987. Subsequently, there are reports from china and Australia. In coming years, the techniques are being invented and the problems of genotype-dependent regeneration will be sorted out.

The first approval for commercial cultivation of Bt cotton in India was granted to three cotton hybrids, MECH-12 Bt, MECH-162 Bt and MECH-184 Bt developed by Mahyco (Maharashtra Hybrid Seed Co.)

Table 3.2. Chronology of Development and Approval of Bt-Cotton in India

Year	*Event*
1995	Mahyco applied to DBT (Department of Biotechnology, Govt. of India) for permission to import a small stock of Bollgard® (*Bt* cotton) seeds from Monsanto Company, USA. DBT gave permission.
1996	A nucleus stock of 100 gms of cotton seeds of the variety Cocker 312 containing the Bollgard® *Bt* gene, *cry 1Ac,* was received by Mahyco from Monsanto, USA.Initiated crossing with the Indian cotton breeding lines to introgress *cry 1Ac* gene.40 elite Indian parental lines were converted for *Bt* trait.
1996-1998	Risk-Assessment Studies conducted using *Bt*-cotton seeds from converted Indian lines.
1998 – 1999	Multi-location field trials at 40 locations in 9 states to assess agronomic benefits and safety. Data submitted to RCGM (Review Committee for Genetic Modification), Ministry of Science and Technology, Govt. of India.
1999 – 2000	Field trials repeated at 10 locations in 6 states. Data submitted to RCGM.

Year	*Event*
2000	July 2000 – Based on the recommendation of RCGM, the GEAC (Genetic Engineering Approval Committee), Ministry of Environment and Forests, Govt. of India, gave approval for Mahyco to conduct large scale field trials in 85 ha and also undertake seed production in 150 ha.
2001	*Kharif* 2001 – Large scale field trials covering 100 ha. Field trials were also conducted by All India Coordinated Cotton Improvement Project of the Indian Council of Agricultural Research (ICAR).
2002	On 26 March 2002, GEAC approved Mahyco's three *Bt*-cotton hybrids, *viz.* MECH 12, MECH 162 and MECH 184, for commercial cultivation in India. This approval was initially valid for three years and also stipulated a few conditions. This is a landmark decision as *Bt*-cotton is the first-ever transgenic crop to receive such a regulatory approval in India.

Source: Manjunath, T. M. (2004).

Figure 3.2. Boll of Bt Cotton.

Table 3.3. Commercial Cultivation of Bt Cotton Undertaken In India – 2002

State	*MECH –12*	*MECH-162*	*MECH – 184*	*Total (Area in hectares)*
Andhra Pradesh	44	5,564	-	5,608
Gujarat	76	4,136	4,642	8,854
Madhya Pradesh	60	404	1,756	2,220
Karnataka	-	3,828	80	3,908
Maharashtra	112	9,300	5,334	14,746
Tamil Nadu	-	2,042	660	2,702
TOTAL	292	25,274	12,472	38,038

Source: Deepak Shaw (2012).

Advantages of Bt-Cotton

- ☆ Good control of bollworm species, the intensity of bollworms reduced significantly on cotton and also on other host crops.
- ☆ Higher boll retention and more yield. Yields are estimated to have increased at least by 30.0 per cent due to effective protection from bollworm damage
- ☆ Reduction in chemical sprays for bollworm control (50 per cent less than that required for conventional commercial hybrids. Prior to the introduction of Bt cotton, about 9400 M tonnes of insecticides were used for bollworm control in India
- ☆ Substantial increase in net income to farmers.
- ☆ No adverse impact on non-target organisms and the adjacent non Bt cotton or other crops.
- ☆ The quality of seed-cotton from Bt-cotton fields was found to be better than non-Bt cotton because of negligible loculi damage and fiber damage.
- ☆ Earliness and determinate habit: Introduction of Bt gene into the hybrids has added the advantage of protection of early fruiting parts, thus resulting in earliness and determinate habit. The earliness ranged from 15 to 20 days in many hybrids in many parts of the country. There have been several added benefits to this. In North India, farmers were able to take up wheat cultivation immediately after early harvest of cotton. The number of picking reduced and the yield per each of the few pickings, increased. Farmers were able to get remunerative returns because of higher prices generally prevalent early in the market during the initial cotton arrivals. (Kranthi, K. R. (2012)).

Bio Safety Issue Related to Bt-Cotton

In India, There are six Competent Authorities to handle various bio-safety issues in transgenic crops

1. Recombinant DNA Advisory Committee (RDAC)
2. Institutional Bio-safety Committee (IBSC)
3. Review Committee on Genetic Manipulation(RCGM)
4. Genetic Engineering Appraisal Committee (GEAC)
5. State Biotechnology Coordination Committee
6. District Level Committee

In general, these authorities are vested with non-overlapping responsibilities.

Conclusion

Cotton is one of the most important fiber crops in india and it plays vital role in Indian economy. The most invasive pest in the cotton is cotton bollworm. And several years bt sprays have been used to control the bollworms. The reason for the introduction of Bt cotton was to counter insect pests *Helicoverpa armigera*, commonly referred to as American Bollworm, which used to cause substantial damage to the crop, resulting in low productivity. Therefore, Mahyco (Maharashtra Hybrid Seed Company), in collaboration with Monsanto, introduced Bt cotton technology into India. Bt cotton hybrids have exhibited excellent control of American Bollworm and reduced the use of insecticides and improved the boll quality and yield. There are several boards are there to regulate bio-safety issues regarding transgenic crops.

REFERENCES

Barwale, R.B., V.R. Gadwal, Usha Zehr and Brent Zher (2004), "Prospects of Bt Cotton Technology in India", *The Journal of Agrobiotechnology Management and Economics*, Vol. 7, Nos. 1 and 2, pp. 23-26.

Karihaloo, J.L. and P.A. Kumar (2009), "Bt Cotton in India: A Status Report", Asia-Pacific Consortium on Agricultural Biotechnology (APCoAB), ICRISAT, NASC Complex, New Delhi.

Khadi B.M., Katageri I.S. and Kulkarni V.N. (2002). Bt Cotton: Basics and its application in India. In '*National Seminar on Bt Cotton*', UAS, Dharwad. 1-25.

Kranthi, K. R. (2012). *Bt cotton-questions and answers*. Indian Society for Cotton Improvement (ISCI), Mumbai

Kumar, P. A., Sharma, R. P. and Malik, V. S. 1996. Insecticidal proteins of Bacillus thuringiensis. *Adv. Appl. Microb.*, 42: 1-43.

Manjunath, T. M. (2004, November).*Bt cotton in India: The technology wins as the controversy wanes*. In: 63rd Plenary Meeting of International Cotton Advisory Committee (ICAC) Meeting, Mumbai (Vol. 28).

Mayee, C. D., Singh, P., Dongre, A. B., Rao, M. R. K., and Raj, S. (2003). Transgenic Bt Cotton.

Milner RJ (1994) History of Bacillus thuringiensis. *Ag. Ecosys. Env.*, 49: 9–13.

Shah, D. K. (2012). Bt cotton in India: a review of adoption, government interventions and investment initiatives. *Indian J. Agril. Econ.*, 67(902-2016-67854).

Stephens SG, Mosley ME (1974) Early domesticated cottons from archaeological sites in central coastal. Peru Am Antiquity 39: 109–122

Zhang BH, Feng R (2000) Cotton resistance to insect and pest-resistant cotton. Chinese Agricultural Science and Technology Press, Beijing).

Chapter 3

Aspects and Applications of Hurdle Technology in Food Preservation

Byreddy Naveena[1]*, Tushar Desale[2], Cheekurthi Pragna[3] and Pranav Pradip Patil[4]

[1]Ph.D., Department of Post-Harvest Processing and Food Engineering, Jawaharlal Nehru Krishi Vishwavidyalaya, Jabalpur, Madhya Pradesh
[2]Assistant Professor, Department of Agricultural Engineering, College of Agriculture, Amalner, Maharastra
[3]Assistant Manager, [4]Production Executive, Spice Sterilization Process, CCL Products Limited (India), Duggirala, Guntur, Andhra Pradesh
[4]Jain Farm Fresh Foods Ltd, Jalgaon, Maharashtra

In recent years, food preservation and food security have become an important factor of concern. In modern era of food processing, the demand for fresh and good quality food products has led to the emergence of hurdle technology. Food industry is increasing its effortsto enhance food safety throughout the world. A variety of conventional and novel techniques have been assessed to solve food safety related problems. Hurdles application in food depends on the nature and chemistry of the food, as well as the processing and storage conditions. Many findings revealed that combination of preservatives at lower concentrations discourage microbial activities more than single preservative at higher concentration. Due to developments in intelligent application of hurdle technology as well as critical factors of food preservation and their interactions, currently hurdle technologies has become as prevalent as it could. Present contribution reviews the brief introduction about hurdle technology, its aspects and applications in different food products for their efficient preservation.

Keywords: *Food security, Hurdle technology, Food safety, Novel techniques, Food preservation.*

Hurdle technology is a process of rendering food to be free from spoilage and pathogenic microorganisms by the combination of one or more preservation methods. The spoilage and pathogenic microorganisms have to pass through these individual approaches called "hurdles" for maintaining their activity in food products (Subha, 2013). More than 50 per cent of the harvested food is often lost due to spoilage in developing countries due to the inadequate technologies for post harvested food preservation. It has been reported by FAO of United Nations that one third of the total harvested food of the world is lost before it is consumed. The preservation of almost all foods is based on combined application of several preservative methods (heating, drying, chilling, curing, acidification, oxygen removal, fermenting, adding preservatives *etc.*). These methods and their underlying principles have been applied since centuries empirically, but are recently applied intelligently using the concept of hurdle technology.

Every food has certain inherent preservation factors such as extent of heat treatment received (F), water activity (aw), low temperature storage (t), redox potential (Eh), pH, *etc.* which may be termed as hurdles, because microorganisms will have to 'jump' these hurdles in order to grow and spoil the product. The stability of the product depends upon the intensity of hurdles present in it. Currently different types of food are producing in the world which categorizes into some groups such as vegetables, fruits, nuts *etc.* The main concern regarding the food is their shelf life and shelf stability.

The main theory of bringing hurdle technology up is severity of applying one preserving factor which can lead to degrade nutritional quality as a result of damages may occur to food substances. To have the wholesome foods accompanied and avoid undesirable changes attributed to non-effective methods (destructive methods), multi-hurdle approaches is considered as an appropriate alternative prior to increasing the intensity of a single hurdle factor *i.e.*, combination of hurdles can emerge more synergistic affections, more reduction of pathogens along with more keeping nutrients. In few recent decades, almost all of food products preserve by hurdle method and the amount of hurdle and the types of them are determined by empirical experiments, however it is still a challenge issue to the methods of assessing food stability (Levine and Slade 1988; Rahman, M.S. 2012).

Hurdle Factors

Addition of sugar to produce jams, heating, deposition of antimicrobial agent from smoking onto food surface and so forth are considered hurdles in traditional cases of food preservation. Since aforementioned factors have flavour- affecting properties contribute in sensory properties of final products. To achieve a good level of shelf life in terms of extending shelf life using hurdle technologies, they must arise at different stages in orders. Depending on the condition of product, a different series of hurdles, to overcome satisfactorily the number of pathogens, adequate level of minimizing the loss of nutrients can be used, but it should be noted that in practice, they can operate simultaneously, synergistically and sequentially.

Need for Hurdle Technology

With the fast growing economy and more and more working women, income amongst middle class has increased and the ongoing trend has been to eat out. Therefore, the demand for processed food has increased in recent times. Moreover, consumers now a days are demanding fresh, natural and minimal processed ready to eat food products.

In hurdle technology, hurdles are deliberately combined to improve the microbial stability and the sensory quality of foods as well as their nutritional quality of foods and economic properties. Thus, hurdle technology aims to improve the total quality of foods by application of an intelligent mix of hurdles (Leistner, 2000).

Table 3.1. Principal Hurdles Used for Food Preservation

Parameter	*Symbol*	*Application*
High temperature	F	Heating
Low temperature	T	Chilling, freezing
Reduced water activity	aw	Drying, Curing, Conserving
Increased acidity	pH	Acid addition or formation
Reduced redox potential	Eh	Removal of oxygen or addition of ascorbate
Bio-preservatives		Competitive ora such as microbial fermentation
Other preservatives		Sorbates, sulfites, nitrites

In dairy industry, use of thermal pasteurisation (TP) is the established food technology for commercial processing of milk. However, degradation of valuable nutrients in milk and its sensory characteristics occurs during TP due to substantial heat exposure. Pulsed electric fields (PEF) and microfiltration (MF) both represent emerging food processing technologies allowing gentle milk preservation at lower temperatures and shorter treatment times for similar, or better, microbial inactivation and shelf stability when applied in a hurdle approach compared to TP. Combining PEF and MF in a hurdle technology for 'cold' pasteurisation of milk proved to be more effective than conventional TP and even higher antimicrobial efficacy of PEF/MF for longer preservation of the microbial milk shelf life can be expected if the processing temperature is increased to a moderate level (≥50 to 65 °C) (Walkling-Ribeiro *et al.*, 2010).

Types of Hurdles

At present, physical non-thermal processes (eg. Pulse electric field) received considerable attention. Another group of hurdles of special interest in industrialized and developing countries at present would be "natural preservatives (spices and their extracts, lysozyme, chitosan, pectin hydrolysate *etc.*). These "green preservatives" are preferred in most of the countries because they are not synthetic chemicals, but in some developing countries, they given preference, since spices are readily available and cheaper than imported chemicals (Leistner, 2011).

1. Physical Hurdles

Most hurdles under this are processes used in the food manufacturing. When using processes intended to kill microorganisms, it is necessary to protect the food product against (microbial) recontamination after processing, this includes: heat processing, manipulation of storage temperatures, irradiation, photodynamic inactivation, ultrahigh pressure processing, ultrasonication, electromagnetic energy, and packaging (Leif, 1994).

2. Physico-chemical Hurdles

Hurdles under this class are; water activity (aw), pH, redox potential (Eh), salt (NaCl), nitrite ($NaNO_2$), carbon dioxide (CO_2), oxygen (O_2), ozone (O_3), organic acids, spices and herbs, sulphite or SO_2, smoking, ethanol, maillard reaction products (MRPs) and lysozyme (Leif, 1994).

3. Microbiologically Derived Hurdles

Hurdles categorised under this heading include competitive flora, starter cultures, bacteriocins and antibiotics (Leif, 1994).

Aspects of Hurdle Technology

The strength or intensity of a hurdle will normally not be sufficient to render the food as safe, but in combination with other hurdles the desired effect can be achieved. Therefore, to ensure the safety and suitability and or to extend the shelf life of milk products, generally more than one hurdle is needed to control microbial content and or growth, to inhibit spoilage and to help prevent food borne diseases. Hurdle Technology is most efficient when it is multi-targeted that is, when various individual hurdles are selected so that different systems of the microorganism are targeted, such as the cell wall, membrane transport, receptor functions, signal transduction, control of gene expression, enzyme system, *etc.* In many cases, a multi-targeted hurdle technology using hurdles with low intensity may be more effective than one single hurdle (treatment or factor) with high intensity.

The hurdle technology affects the physiology and growth of microorganisms in food. There are mainly 4 major mechanisms by which hurdle technology affects the growth of microorganisms in foods:

Homeostasis

It is the process that maintains the stability of the living cells internal environment in response to the changes in external environment. In food preservation the homeostasis of microorganisms is a key phenomenon, because if the homeostasis of these microorganisms is disturbed by preservative factors (hurdles) in foods, they will not multiply, *i.e.* they remain in the lag-phase or even die, before homeostasis is repaired (re-established). Therefore, food preservation is achieved by disturbing the homeostasis of microorganisms in a food temporarily or permanently. Some of the examples of homeostasis in the body are regulation of temperature and balance between acidity and alkalinity (pH). These factors are prerequisite feature of living cells and this applies to microorganisms as well as higher organisms (Leistner, 2000). The concept behind homeostasis is already known in higher organisms but this knowledge should be incorporated in microorganisms important for the poisoning and spoilage of foods (Leistner, 2000). Disturbing the homeostasis of the microorganisms by various hurdles eventually results in the death of the spoilage causing microbes thereby protecting the food product from microbial spoilage.

Metabolic Exhaustion

This aspect of hurdle technology deals with auto sterilization of food. This was firstly observed in the experiment carried out on liver sausages inoculated with Clostridium sporogens and stored at 37 °C. Later this behaviour of some bacterial spores was regularly observed in shelf stable meat products during storage time period (Leistner, 1994b).

The most likely explanation is that bacterial spores which survive the heat treatment are able to germinate in these foods under less favourable conditions than those under which vegetative bacteria are able to multiply. Thus, the spore counts in stable hurdle-technology foods actually decrease during storage of the products, especially in unrefrigerated foods.

It was observed that in hurdle technology treated food products the spore counts decreases during storage especially at ambient temperature. The microbes in the hurdle treated stable products uses their energy for homeostasis thereby becoming metabolically exhausted. This leads to auto sterilization of food products. Thus, the microbiologically stable product becomes safer for storage at normal room temperatures.

Multi Target Preservation of Food

It is a very important aspect for efficient and effective preservation of targeted food material. Hurdles applied in the targeted food material might not just have effects on microbial stability but they could act synergistically. Synergistic effect could be achieved in the targeted food, if the hurdles affects different targets such as pH, aw, enzyme systems simultaneously within the microbial cell and thus disturb the homeostasis of the microbes rendering it difficult for the microbes to synthesise different stress shock proteins and to maintain their homeostasis (Leistner, 1995a). Therefore the application of several hurdles simultaneously would lead to an optimal microbial stability and effective food preservation.

Applications of Hurdle Technology in different Products

Hurdle Technology is a novel concept which has several applications in the preservation of various food products such as:

Dairy Products

To enhance the shelf life of dairy products, Hurdle technology has been applied in many dairy products. By applying various hurdles such as pH, a_w, preservatives and Modified Atmosphere Packaging (MAP) shelf stable paneer can be prepared. The quality and shelf life of hurdle treated paneer extended from one to twelve days at ambient temperature and six to twenty days at refrigeration temperature without affecting its physicochemical and sensory properties (Thippeswamy *et al.*, 2011).

The shelf life of paneer curry can be increased using hurdle technology. Compared to heat sterilized product, the hurdle treated paneer will have better quality (Rao and Patil, 1999). Another product, a traditional Indian heat desiccated milk khoa based product brown peda have also been prepared and preserved through hurdle technology. Hurdle treated brown peda could be best preserved without any quality loss up to 40 days at room temperature (30±1°C) (Panjagari *et al*, 2007).

Gasasase (poppy seeds) payasum is product prepared from poppy seeds and rice and by application of retort processing (f0 value 6 at steam pressure of 1.04 bar) to this product, there was a marginal change in its pH, acidity and HMF content and viscosity take place during storage which led to extension of its shelf-life to 6 weeks at 37°C temperature (Geetha, 2005).

Dudhchurpi, a product of Himalaya region (Bhutan, Sikkim and Darjeeling) made from milk of yak or cow and is self-stable for several months without refrigeration. Most important for dudhchurpi is its texture (elasticity), since people living at high altitude chew it as 'energy tablets'. Sensory analysis and microbial stability of dudhchurpi was optimized by Hossain using combined several hurdles like heating, acid coagulation, addition of sugar and sorbate, smoking, drying and packaging in a closed container (Hossain, 1994).

Fruits and Vegetables

In the preservation of various fruits and vegetables like carrot, pineapple, coconut and papaya, several hurdles are considered to be important to enhance their stability and shelf life. Shelf-stable grated carrot products are developed using hurdle technology. Hurdle technology can also be applied to develop shelf stable RTE (Ready-To-Eat) intermediate moisture pineapple with increased shelf life. Osmotic dehydration, infrared drying and gamma radiation can successfully reduce the microbial load in pineapple slices increasing its shelf life up to 40 days (Saxena *et al.*, 2009).

Fruit Derived Products

For the preservation of various fruit products, several conventional hurdle strategies are effectively used along with the novel ones. UV light, pulsed light (PL), ultrasound (US), and high hydrostatic pressure (HHP) are some of the hurdles applied in fruit processing. Several potential hurdles like heat treatment, preservatives, irradiation and various packaging materials can be used for the preservation of Sugarcane juice (Sankhla *et al.*, 2012).

The applicability of these hurdles shows great enhancement in the level of product safety and stability and thus it can therefore be recommended for preservation of all kinds of food material (Sankhla *et al.*, 2012). Hurdle or combined technology is also applied in the preservation of high moisture fruit products such as peach, pineapple, papaya, mango and banana. The technology is based on the combination of heat treatment, water activity and addition of antimicrobials (Alzamora *et al.*, 1993).

Meat and Meat Products

Hurdle technology has been applied to a number of meat products. The combined effect of the hurdles on pork sausages resulted in the inhibition of the growth of yeast and moulds up to 12 days, while the dipping of sausages into potassium sorbate solution before packaging inhibited their growth up to 30 days. Implementation of hurdle technology to sausages was beneficial as the shelf life of sausages was increased up to 30 days while the control untreated samples were acceptable up to 18 days (Thomas *et al.*, 2010). Shelf stable ready to eat pickle type spiced buffalo meat products was also prepared and preserved by controlling different hurdles like pH, water activity, proximate composition, FFA, Soluble hydroxyproline, TBA values, nitrite content, protein solubility (Malik and Sharma, 2014). Hurdle Technology has made it possible to deliver shelf stable and ready to eat meat products for space scientists and mountaineers.

Conclusion

Hence, hurdle technology is used in industrialized as well as in developing countries for the effective preservation of foods by which a preservation parameter can be used at an optimum level in order to get a maximum lethality against microorganisms by a combination of 2 or more such parameters, so that damage to sensory parameters of food is kept to minimum. The adverse effect of some technologies on the sensory, texture and nutritional values of food could be overcome by using hurdle technology. The hurdle technology makes minimal sensory and nutritional changes in the product which makes the product more valuable and acceptable than obtained by conventional methods and has become a boon for the efficient preservation of food products.

REFERENCES

Alzamora, S.M., Tapia, M.S., Argaíz, A., Welli, J. (1993). Application of Combined Methods Technology in Minimally Processed Fruits. *Food Res. Int.* 26, 125 130. Doi: 10.1016/0963-9969(93)90068-T.

Geetha P. (2005). Studies on physico-chemical and sensory characteristics of retort processed gasagase payasam. M.Tech. Thesis submitted to National Dairy Research Institute (Southern Campus), Bangalore.

Hossain SKA. (1994). Technological Innovation in Manufacturing DudhChurpi. Ph.D. Thesis, University of North Bengal, Siliguri, India, pp.122.

Leif B. (1994). Discription of Hurdles. In Leistner L. and Gorris L. G.M. (Eds). Food Preservation by Combined Processes. European Commission, Final Report on Food-Linked Agro-Industrial Research (FLAIR) Concerted Action No. 7, Subgroup B. Pp 8-25.

Leistner, L. (1995a). Principles and applications of hurdle technology. In: Gould, G.W. (Ed.), New *Methods for Food Preservation,* Blackie Academic and Professional, London, pp. 1-21.

Leistner, L. (2000). Basic Aspects of Food Preservation by Hurdle Technology. *Int. J. Food Microbiol.*, 55, 181 186. Doi: 10.1016/S0168-1605(00)00161- 6.

Leistner L. (2011). Hurdle Technology. Encyclopedia of Life Support Systems (EOLSS), Food Engineering-Vol III.

Leistner, L. (1994b). Food Design by Hurdle Technology and HACCP, Printed and distributed by the Adalbert- Raps-Foundation, Kulmbach, Germany, p. 62.

Levine, H. and Slade, L. (1988). Water as a plasticizer: physico-chemical aspects of low-moisture. Polymeric systems. In: Franks F (ed) *Water Science Reviews*, Vol. 3. Cambridge University Press, Cambridge.

Malik, A.H. and Sharma, B.D. (2014). Shelf Life Study of Hurdle Treated Ready-to-Eat Spiced Buffalo Meat Product Stored at 30 ± 3 °C for 7 Weeks Under Vacuum and Aerobic Packaging. *J. Food Sci. Technol.*, 51, 832 844. Doi: 10.1007/ S13197-011- 0592-9.

Panjagari, N.R., Londhe, G.K. and Pal, D. (2007). Effect of packaging techniques on self-life of brown peda, a milk based confection. *J. Food Sci and Tech.*, 47: 117-125.

Rahman, M.S. (2012). Applications of macro-micro region concept in the state diagram and critical temperature concepts in determining the food stability. *Food Chemistry*, 132: 1679–1685.

Rao, K.J. and Patil, G.R. (1999). Development of ready to eat paneer curry by hurdle technology. *J. Food Sci. Tech.*, 36: 37- 41.

Sankhla, S., Chaturvedi, A., Kuna, A. and Dhanlakshmi, K. (2012). Preservation of sugarcane juice using hurdle technology. *Sugar Tech.*, 14: 26 39. Doi: 10.1007/ S12355-011-0127-8.

Saxena, S., Mishra, B.B., Chander, R. and Sharma, A. (2009). Shelf stable intermediate moisture pineapple (*Ananas comosus*) slices using hurdle technology. *LWT - Food Sci. Technol.*, 42, 1681 1687. Doi: 10.1016/J.Lwt.2009.05.009.

Subha, G. (2013). Biologically viable methods for food preservation: A review. *Res. J. Chem. Env. Sci.*, 1(2): 1-2.

Thippeswamy, L., Venkateshaiah, B.V. and Patil, S. (2011). Effect of modified atmospheric packaging on the shelf stability of paneer prepared by adopting hurdle technology. *J. Food Sci. Tech.*, 48: 230-235. Doi: 10.1007/S13197-010-0155-5.

Thomas, R., Anjaneyulu, A.S.R. and Kondaiah, N. (2010). Quality of hurdle treated pork sausages during refrigerated (4 ± 1°C) storage. *J. Food Sci. Tech.*, 47: 266-272. Doi: 10.1007/S13197-010-0041-1.

Chapter 4

Present Scenario of Soil Testing and Fertilizer Recommendations in Odisha

Antaryami Mishra[1], Subhashis Saren[1], Prava Kiran Dash[1] and Pradip Dey[2]*

[1]Odisha University of Agriculture and Technology, Bhubaneswar, Odisha
[2]ICAR-Indian Institute of Soil Science, Bhopal, Madhya Pradesh

The state of Odisha covering a geographical area of 15.56 mha has eight broad soil groups (genetic approach) such as Red soil (7.14 mha), Laterite soil (0.70 mha), Red and Yellow soil (5.50 mha), Black soil (0.96 mha), Alluvial soil (0.67 mha), Mixed Red and Black Soil (0.16 mha), Brown forest soil (0.17 mha) and saline soil (0.26 mha) distributed in 30 districts. Taxonomically, the state comprises four major soil orders viz. Inceptisol (49 per cent), Alfisol (35 per cent), Entisol (10 per cent) and Vertisol (6 per cent) (Sahu and Mishra,2005). The cultivable area is around 6.70 mha. As far as soil reaction is concerned on a district wise basis 69 per cent soils of Odisha are acidic, 6 per cent saline and rest 25 per cent are neutral; 60 per cent soils are medium and 40 per cent soils are low in organic carbon content; 73 per cent are medium and 27 per cent soils are low in available phosphorus. With respect to available potassium 7 per cent soils are low, 86 per cent are medium and 7 per cent are high. The soils of Ganjam and Gajapati districts are low and Boudh and Kandhamal districts are high in available soil potassium content.

Keywords: *Soil testing, Fertilizer recommendations, Soil orders, Organic carbon, Phosphorous, Potassium.*

Soil Testing

To cope with the green revolution of the country in first and second five year plan, eleven soil testing laboratories were established in 11 districts (old) of Odisha for soil testing services with minimum capacity of 10,000 soils samples per annum in each laboratory. Later, seven new laboratories have been established. Establishment of six more laboratories are in the pipeline. As far as OUAT is concerned 11 soil testing laboratories are functioning in 11 different KVKs. Six more STL is going to be established in 6 more KVKs in the coming financial year. The soils are analyzed for pH, organic carbon; available N, P and K and micronutrients with standard procedures and recommendations are made by the Soil Chemist, OIC of Soil Testing Laboratories for different crops in Odisha. Beside this, blanket recommendations are also practiced by the state Department of Agriculture.

Colwell (1971) rightly defines a Soil Test – "A measurement qualifies to be termed as soil test for a particular nutrient if it provides information on pH, texture, depth of sub-soil, availability of moisture *etc.* besides the amount of extractable nutrients, those have significant effect on nutrient entry to plant roots"

Soil Test Calibration and Fertilizer Recommendation

The biological relevance of Soil Test calibration is to establish relationship between the level of soil nutrients as determined in the laboratory and the crop response to fertilization observed in the fields. The complexity arises due to diversity of soils, climate, crop variety and management practices. In spite of these complexities, successful attempts have been made in making use of soils testing as a predictive tool for fertilizer recommendations. The soil test values for organic carbon, available N, P and K estimated by appropriate methods are categorized under three or five fertility levels (very low, low, medium, high and very high). Soil test value rated low indicates high probability of profitable response to applied fertilizer and vice-versa (Muhr *et al.*, 1965).

The generalized fertilizer recommendation is equated with the soils rated medium and for soils rated under high or low category; the generalized recommended dose is decreased or increased by 25 per cent for field crops.

The following soil test ratings are adopted by the soil testing laboratories of Odisha.

Sl. No.	*Nutrient/soil parameters (method)*	*Rating*				
		Very Low	*Low*	*Medium*	*High*	*Very High*
1.	Organic carbon per cent (Wet digestion method)	-	<0.5	0.5-0.75	>0.75	-
2.	Available N (kg ha^{-1}) (Alkaline permanganate)		<250	250-500	>500	
3.	Available P (kg ha^{-1})					
	a. Olsen's	<4	4-9	9-22	22-35	>35
	b. Bray's	<6	6-14	14-40	>40	
4.	Available K (kg ha^{-1}) (Neutral Normal Ammonium Acetate)	-	<118	118-280	>280	

Conversion factor: P to P_2O_5=2.3 and K to K_2O= 1.2.

Adjustment of Fertilizer Dose

Requirement of the Crop for Nutrient	*Soil Test Rating*	*± Over the Generalized Dose*
High	High	-25 per cent
High	Low	+35 per cent
Low	High	-35 per cent
Low	Low	+25 per cent

This calibration approach is somewhat qualitative. Some of the short-falls are-

1. Ratings of organic carbon and $KMNO_4$-N are not equated and rating is based on organic carbon.
2. NH_4OAc – K does not reflect the status properly for light to medium textured soils. 1*N* HNO_3 – K is better.
3. When the test values of P and K are very low, doses may be raised by 50-100 per cent to take care of the fixation.

Use of Nutrient Indices and Fertility Maps

The Project has prepared block level GPS and GIS based soil fertility maps for nine districts of Odisha *viz.* Puri, Cuttack, Khurda, Nayagarh, Dhenkanal, Angul, Sambalpur, Bhadrak and Balasore comprising of 87 blocks. Fertility maps include (pH, EC, Organic carbon, available Nitrogen, Phosphorus, Potassium, Sulphur and Boron). These maps are very much helpful in site specific nutrient management and will be helpful for online fertilizer recommendation to the farmers for different crops of Odisha.

Since it is not possible for soil testing of the soil of every farmer's field every year, area wise (Block or District) soil fertility maps are to be prepared using Nutrient indices. The generalized recommendation for various crops may be refined by adjustment depending upon the Nutrient indices of particular area. It is possible to relate the Nutrient indices to major soil types, major cropping system or agro-climatic situation and interpolate the recommendation. District wise nutrient indices have been prepared for the state.

Fertilizer Recommendation on Yield Targets

Liebig's Law of Minimum forms the basis of fertilizer recommendation for yield targets first advocated by Truog (1960). Ramamoorthy *et al*, (1967) established the critical basis and experimental proof of this approach of soil test calibration through soil test crop response (STCR) studies creating fertility gradients in a large experimental field rather than conducting multi-location field trials to avoid variations in crop management effects. The parameters taken into account for formulation of fertilizer prescription equations for targeted yield of crops.

1. Nutrient requirement (NR) kg q^{-1} of economic product
2. Efficiency of soil estimated nutrient (per cent)=Cs

3. Efficiency of fertilizer nutrient (per cent)= Cf
4. Efficiency of organic matter (per cent)= Co

$$FD\,(kg\,ha^{-1}) = \frac{NR \times 100T}{Cf} - \frac{Cs}{Cf} \times SN \text{ or } SP \text{ or } SK$$

Where, T is yield target in q ha^{-1} and S is the soil estimated nutrient kg ha^{-1}

In subsequent period, attempts have been made to extend this approach of calibration for area wise fertilizer recommendations based on yield targeting and nutrient index (Ramamoorthy *et al.*, 1969), fertilizer recommendation for crops in multi-cropping systems (Randhawa and Velayutham, 1982) and integrated nutrient management using both inorganic and organic sources for yield targets (Rani Perumal *et al.*, 1984).

The STCR project in Odisha was started in the year of 1996. Work in these lines have been initiated for generation of required information as regard to optimum yield targets, efficiency of soil, fertilizer and manurial sources of nutrients and nutrient requirement of principal crops of Odisha under varying soil and agro-climatic conditions. Since inception of the project fertilizer prescription equations for targeted yield of 15 crops have been formulated. The crops are Rice (Mishra *et al.*, 2009), Groundnut (Pradhan *et al.*, 2007), Sesamum (Mishra *et al.*, 2008), Pumpkin (Gogoi *et al*, 2011), Lady's finger (Mishra *et al.*, 2013), Brinjal (Mishra *et al.*, 2013), Potato, Tomato (Mishra *et al.*, 2013), Mustard (Mishra *et al.*, 2011), Chilli (Mishra *et al.*, 2014), Cabbage (Saren *et al.*, 2015), French bean (Pogula *et al.*, 2016), Black gram (Saren *et al.*, 2017), Cowpea (Mishra *et al.*, 2017),Green gram (Saren *et al.*, 2018), Potato, Knolkhol, Pea, Bottle gourd, Bitter gourd and Carrot. The validity of these equations have been verified in more than300 nos. of farmers' field through frontline demonstration (FLD) covering six agro-climatic zones out of the total of ten agro-climatic zones of Odisha. The results have shown superiority of STCR developed and recommended technology over farmers' practice and soil test based recommendation *i.e.* on the basis of low-medium-high (LMH).

Conclusion

The calibration approach appears to be the most scientific and quantitative, we have to go a long way in refining the estimation of various parameters under diversified conditions, more particularly for problematic situations (acid soils, saline alkali soils, dry lands, water logged conditions *etc.*) Further deficiencies and interaction of secondary and micro-nutrients have to be taken into consideration to arrive at a well balanced and profitable recommendation. Yield targets have to be dynamic depending on the variation of the unit cost of nutrients and produce.

REFERENCES

Colwell, J.D. (1971). Porc. Int. Symp. *Soil Feritlity Evaluation*, Vol.1, 327.

Gogoi, A., Mishra, A. and Jena, B.(2011). Soil Test based fertilizer recommendation for targeted yield of pumpkin (*Cucurbita moschata*) under rice-pumpkin cropping system in an *Inceptisol* of Odisha. *Environment and Ecology*, 29(2): 574-576.

Mishra, A., Das, D. and Saren, S. (2013). Soil test based fertilizer recommendation for targeted yield of Brinjal (*Solanum melongena*) under Rice-Brinjal cropping system in an *Inceptisol* of Odisha. *Odisha Horticulture,* 2(2): 62-69.

Mishra, A., Dash, B.B, Nanda, S.K., Das, D. and Dey, P. (2013). Soil Test Based Fertilizer Recommendation for Targeted Yield of Tomato (*Lycopersicon esculentum*) under Rice-Tomato Cropping System in an *Ustochrept* of Odisha. *Environment and Ecology,* 31(2A): 655-658.

Mishra, A., Dash, B.B., Nanda, S.K. and Das, D. (2013). Soil Test Based Fertilizer Recommendation for Targeted Yield of Lady's Finger (*Abelmoschus esculentus*) under Rice-Lady's Finger Cropping System in an *Ustochrept* of Odisha. *Environment and Ecology,* 31(1): 58-61.

Mishra, A., Dash, B.B., Nanda, S.K., Das, D., Sarangi, J., Panda, N. and Mishra, H.T.(2011). Study of Soil Taxonomy as a basis for extrapolation of fertilizer prescription equations for targeted yield of crops to different areas- a case study in a *Vertic Ustochrept* of Odisha. *J. of Research,* 1(1): 207-214.

Mishra, A., Nanda, S.K., Pradhan, N.K., Jena, B. and Mukhi, S.K. (2009). Soil test based fertilizer recommendation for targeted yield of rice (*Oryza sativa*) under rice-rice cropping system in an *Inceptisol* of Odisha. *Extended summaries-Symposium on resource management in corps and cropping systems under changing climate,* ISA and OUAT, Bhubaneswar, 7-8May, 2009.

Mishra, A., Pradhan, N.K., Nanda, S.K and Jena, B. (2008). Soil Test Based Fertilizer Recommendation for Targeted Yield of Sesamum (*Sesamum indicum*) under Rice-Sesamum Cropping System in an *Inceptisol* of Odisha. *Environment and Ecology,* 26(4A): 1756-1758.

Mishra, A., Saren, S. and Dey, P. (2016). Integrated Nutrient Management and Formulation of Targeted Yield Equations for Cowpea [*Vigna unguiculata* (L.) Walp.]. *Environment and Ecology,* 35 (2A): 867—872.

Mishra, A., Saren, S., Das, D and Dey, P. (2014). Optimization of chilli yield under rice- chilli cropping system in an *Inceptisol* of Odisha. *Indian Agriculturist, 58 (1): 19-24.*

Muhr, G.R., Datta, N.P., Subramoney, H., Leley, V.K. and Donahue, R.L. (1965). Soil Testing in Indis, USAID, New Delhi.

Pogula, S., Truptimayee, Mishra, A., Saren, S. and. Dey, P. (2016). Soil Test Based Fertilizer Recommendation for Targeted Yield of French Bean (*Phaseolus vulgaris*) under Rice-French Bean Cropping System. *International Journal of Bio-resource and Stress Management,* 7(5): 1128-1130.

Pradhan, N.K., Mishra, A., Padhy, G.P. and Jena, B. (2007). Soil test based fertilizer recommendation for targeted yield of groundnut (*Arachis hypogaea*) under Rice-Groundnut cropping system in an *Inceptisol* of Odisha. *Environment and Ecology,*25 (2): 478-480.

Ramamoorthy, B., Bajaj, J.C. and Singh, K.D. (1969). *Fert. News,* 14 (10): 45-51.

Ramamoorthy, B., Narasimhan, R.L. and Dinesh, R.S. (1967). *Indian Farming* 17(5): 43-45.

Randhawa, N.S. and Velayuthum, M. (1982). *Fert. News.* 27 (9): 35-64.

Raniperumal, Manikam, T.S., Ramanthan, P. and Mani, S. (1984). *Tamil Nadu Agriculture Journal*, Tamil Nadu Agril. University, Coimbatore, pp. 1-36.

Sahu, G.C. and Mishra, A. (2005). Soils of Odisha and their Management. *Odisha Review.* LXII (4): 56-60.

Saren, S, Mishra A. and Dey P. (2018). Site specific nutrient management through targeted yield equations formulated for green gram (*Vigna radiata* L. Wilczek). *Legume Research*, 41 (3): 436-440.

Saren, S, Mishra, A. and Dey, P. (2017). Integrated Nutrient Management and Formulation of Targeted Yield Equation for Blackgram (*Vigna mungo* L.). *Current Science*, 113 (2): 314-317.

Saren, S., Mishra, A, Das, D. and Dey, P. (2015) Soil Test Based Fertilizer Recommendation for Targeted Yield of Cabbage under Rice-Cabbage Cropping System in an Inceptisol. *Indian Agriculturist*, 59 (1): 55-60.

Truog, E. (1960). Trans. 7th Int. Cong. Soil Sci. Vol. VIII, Commission, IV Paper No. 7, pp. 46-53.p.

Chapter 5

Growth Rates in Area Production and Yield of Pulses in Meghalaya: A Decomposition Analysis

N.R. Meena

Assistant Professor, College of Agriculture, NDUAT, Kumarganj, Ayodhya, Uttar Pradesh

The present study evaluated the contribution of area, yield and their interaction to growth in production of pulses as well as the compound annual growth rate in area, production and yield of pulses in seven districts of Meghalaya. Time series data on area, production and yield of pulse for the period 2005-2017 were collected from secondary sources.The decomposition analysis of growth suggested that the major contribution in the change of pulse production in the Meghalaya was interaction effect (47.74 per cent) followed by yield effect (35.68 per cent) and area effect (16.56 per cent).Among the 7 districts of Meghalaya, East Garo Hills registered highest growth in area and production followed by East Khasi Hill. While, East Khasi Hills district registered highest positive growth rates in yield increased followed by East Garo Hills and least growth rate was observed in West Garo Hills under the crop over the time period of 13 years (2005-2017).

***Keywords**: Pulse, Meghalaya, Growth rate, Decomposition analysis.*

Pulses are one of the important food crops in the world particularly for poor farmers and low income family and a major source of protein for vegetarian masses. Pulses are rich source of protein (20 – 25 per cent) along with other nutrients like carbohydrates, fiber, calcium, iron, zinc, folic acid essential vitamins and other minerals. These crops are grown in marginal land or grow as inter crop with cereals and oilseed crops. They produce income for farmer and consumption of pulses help to resolve fatness, cancers, cholesterol, diabetes, heart disease and malnutrition (GoI, 2015). The by- products of pulses like leaves, pod coats and bran are also uses as dry fodder for animal, while gram, lobia, urd bean and moong bean are used as green fodder to feed the animal (Joshi and Rao, 2017). The world acreage under pulses was reported about 85.40 million ha with production of 87.40 metric tonnes and productivity of 1023 kg/ha in 2017. India ranked first in area (29.99 Mha) under pulses in 2017 followed by, Niger (54.7 Mha), Myanmar (42.03 Mha), Nigeria (38.49 Mha), Brazil (32.09 Mha), Canada (28.7 Mha), China (23.85 Mha) and Tanzania (20.68 Mha). Regarding the production of pulses in world, India (199.8 mt) ranked first followed by Myanmar (59.77 mt), Canada (58.28 mt), China (41.43 mt), Ethiopia (26.13 mt) and Australlia (22.47 mt), Although, India ranked first in area and production in world under pulses but its productivity is very low. The productivity of pulses in India was 841kg/ha that is far below to world average productivity (909 kg/ha) in 2017-18. The productivity of pulse was increased from 524 kg/ha in 1970-71 to 841 kg/ha during 2017-18 (FAO, 2017).

Regarding area under total pulses, Madhya Pradesh (7480 thousand ha) ranked first, followed by Rajasthan (5329 thousand ha), Maharashtra (4352 thousand ha) and Uttar Pradesh (2268 thousand ha). Similarly, in terms of production also Madhya Pradesh (8112 thousand tonnes) secured first ranked followed by Rajasthan (3386 thousand tonnes), Maharashtra (3304 thousand tonnes) and Uttar Pradesh (2208 thousand tonnes). The yield of pulses was highest in Madhya Pradesh followed by Uttar Pradesh, Maharashtra and Rajasthan (FAO, 2017). Chickpea cover the highest area under group of pulses grown in India followed by Urad bean, pigeon pea and mung bean. Among the pulses, chickpea is the single largest category of pulses contributing 44.5 per cent to the basket of total pulses followed by pigeonpea and uradbean during 2017-18. Madhya Pradesh ranked first in both area and production of chickpea, lentil and Urad bean during 2017-18. The state of Maharashtra has secured first position in terms of area and production of pigeon pea, while Rajasthan is the largest producer producer of Mung bean. The area under pulses has increased by 6.5 thousand hectares from 2.6 thousand hectares in 2005-06 to 9.2 thousand hectares during 2017-18 in Meghalaya. On the contrary, production of pulses also registered an increase of 8.08 thousand metric tonnes from 4.25 thousand metric tonnes in 2005-06 to nearly 13.14 thousand metric tonnes in 2017-18 (GoI, 2017). Although India is the largest producer of pulses in the world, its domestic demand is still not completed by domestic production. The shortfall is met from imports. The long term solution for meet demand for pulses needs to increase area and productivity of pulses in the country. Area and productivity could increase with the application of modern biotechnologicaltools, trained man power and adoption of innovative technology like short duration and drought resistant variety, use of quality inputs, bring more area under pulses with the help of mixed cropping, crop rotation and intercropping with cereals and millets. Which reduce the poverty, hunger and malnutrition problem and improve the economic status of farmers. Availability of this backdrop, the present study seeks to examine the important aspects related to growth of pulses from a broad perspective in the Meghalaya state. The findings would be helpful background for researcher and policymakers that would help for them to taking decision relating to the pulses.

Data and Methodology

To meet the objective of the study, Secondary data on area, production and productivity of pulses were collected from Directorate of agriculture Govt. of Meghalaya, for a period of 13 years from 2005-06 to 2017-18, to analyze the growth rate with its test of significance and decomposition of growth components. The data were collected for all the districts. The state has 11 districts but data analysis involved 7 districts only and four new districts were curved recently, therefore data on area production and yield were available only for seven districts namely South Garo Hills, West Garo hills, East Garo Hills, Jaintia Hills, East Khasi hills, Ri-Bhoi and West Khasi hills.

Growth Rate Estimation

The average annual compound growth rates of area, production and yield were worked out using the formula recommended by Dandekar (1980). The estimation of compound growth rate seems to be more appreciable to analyze the movement of agricultural output rather than linear rate of growth (Dandekar, 1980). The Compound Annual Growth Rates (CAGR) was worked out by fitting exponential function of the following type to the data for the periods explained above.

$$\log Y = a + b^t$$

$$Log\ Y = \log a + \log b$$

$$CGR = (\text{Antilog } b\text{-}1) \times 100$$

Where, Y is the dependent variable (area, productivity, production) and't' is the time period in years.

Decomposition Analysis

The sources of production growth was examined by using the decomposition model proposed by Sharma (1977) and relative contribution of area, productivity and their interaction effect to the total output change in pulses was worked out as follows:

$$P_n - P_0 = A_0(Y_n - Y_0) + Y_0(A_t - A_0) + (A_t - A_0)(Y_t - Y_0)$$

$$\Delta P \quad = \quad A_0(\Delta Y) \; + \; Y_0(\Delta A) \quad + \quad (\Delta A\ \Delta Y)$$

Change in production = Yield effect + Area effect + Interaction effect

Where, P_n, Y_n and A_n are the production, yield and area under pulses in the current periods respectively and P_0, Y_0 and A_0 are the production, yield and area under pulses in the base periods respectively.However, this method has a flaw in that it only considers the discrete at a and doesnot take all the years into consideration. So, another method for decomposition analysis suggested by Palanisami, Paramisivam and Ranganathan (2002) wastried:

$$P_t - P_0 = A_0(Y_t - Y_0) + Y_0(A_t - A_0) + (A_t - A_0)(Y_t - Y_0)$$

$$\Delta P = A_0 \Delta Y_t + Y_0 \Delta A_t + \Delta A \, \Delta Y_t$$

Change in production = Yield effect + Area effect + Interaction effect

Where, P, A and Y represents production, area and yield. '0' and't' represents time periods such that't'> '0' by an accounting period, which is usually a single year. The result obtained from this method is more reliable as the analysis is done for the continuous data.

Results and Discussion

Compund Annual Growth Rate

To estimate the growth performance of area, production and yield of pulses in the district wise time series data on area, production and productivity of pulse for the period 2005-06 to 2017-18 were analyzed. In this period, the growth in area was observed highest in East Garo Hills (23.08 per cent), followed by West Khasi Hills (18.31 per cent), West Garo Hills (16.50 per cent) South Garo Hills (15.51 per cent) and least were observed in Ri-Bhoi (6.08 per cent) and these were positively significant. It may be due to pulses started to grow as inter crop, mixed crop and involved in intercropping with other cereals and millets. The same trend was observed in production growth highest in East Garo Hills (32.51 per cent), followed by West Khasi Hills (30.05 per cent), East Khasi Hills (26.74 per cent) and South Garo Hills (24.95 per cent) and all were positively significant. In respect to productivity it was highest in East Khasi hills (12.33 per cent) followed by West Khasi Hills (9.92 per cent), Ri-Bhoi (8.36 per cent), South Garo Hills (8.19 per cent), East Garo hills (7.03 per cent) and West Garo Hills (5.80) per cent per annum. All the district of Meghalaya have shown positive growth rate in area, production and yield of pulse during the study period *i.e.* 2005-06 to 2017-18.

Table 5.1. Annual Compound Growth Rates in Area, Production and Yield of Pulses

Districts	*Area*	*Production*	*Yield*
Ri- Bhoi	6.08	14.96	8.36
East Khasi Hills	12.82	26.74	12.33
West Khasi Hills	18.31	30.05	9.92
Jaintia Hills	6.87	12.89	5.36
East Garo Hills	23.08	32.51	7.03
West Garo Hills	16.50	23.27	5.80
South Garo Hills	15.51	24.98	8.19
Meghalaya	16.52	33.88	8.45

It was due to the efforts of ICAR and CAU (I) that started the breeder seed production of HYV_S and their multiplication and distribution among the farmers involving SHG_S in co-ordination with NSC and different governments and other facility such as modern or innovative inputs and requisite training make available to farmers by ICAR NEH region and CAU.

Decomposition Analysis

The growth analysis (area, production and yield) of pulses revealed the general pattern of growth and direction of changes in yield and area. But this analysis does not evaluate the contribution of area and yield towards the production growth so, it is necessary to examine the sources of output growth. To appraise the sources of output growth, the change in production is divided in to three effects *i.e.*, area effect, yield effect and interaction effect.

Table 5.2. Contribution of Area, Yield and their Interaction in the Production of Pulses

Districts	ΔP	$Y_0\Delta A$	$A_0\Delta Y$	$\Delta A\Delta Y$	*Per cent Contribution*		
					Area	*Yield*	*Interaction*
Ri- Bhoi	91	40.88	23.11	27.00	44.93	25.39	29.67
East Khasi Hills	2231	630	380	1210	28.38	17.11	54.49
West Khasi Hills	240	29	93	116	12.40	39.09	48.50
Jaintia Hills	167	58	62	46	35.07	37.43	27.49
East Garo Hills	1600	123	741	735	7.74	46.31	45.93
West Garo Hills	6211	779	3193	2237	12.55	51.41	36.02
South Garo Hills	901	156	349	396	17.36	38.72	43.90
Meghalaya	13574	2249	4844	6482	16.56	35.68	47.74

With the help of this additive decomposition model the relative contribution of area, productivity and their interaction on pulses production of seven districts in the Meghalaya states of north east for period (2005-06 to 2017-18) have been estimated and presented in Table 5.2. Table also reveals that during the study period(2005-06 to 2017-18) the major contribution in the pulse production in Ri-Bhoi district was found due to positive area effect 44.93 per cent followed by yield effect 29.67 per cent and least contribution was due to interaction effect (25.39 per cent). In East Khasi Hills, the interaction effect was found to be highest 54.49 per cent followed by yield effect 28.38 per cent and area effect was only 17.11 per cent. In the West Khasi Hills, interaction effect has highest contribution 48.50 per cent in pulse production followed by area effect 39.09 per cent. In Jaintia Hills, area effect has major contribution 37.43 per cent in pulse production followed by yield effect 35.07 per cent and interaction effect 27.49 per cent. In the West Garo Hills, major contribution was found to be area effect 46.31 per cent followed by interaction effect 45.93 per cent and yield effect 7.74 per cent. In South Garo Hills, interaction effect has the maximum contribution in pulse production followed by area effect 38.72 per cent and yield effect 17.36 per cent. Similarly, in the Meghalaya state also major contribution in pulse production was due to interaction effect 47.74 per cent followed by area 35.68 per cent and yield effect was 16.56 per cent. Similar, results were reported by Meena *et al.* (2017), Roy *et al.* (2015), and Roy *et al.* (2014).

Conclusions

In the Meghalaya, over the period from 2005 to 2017 there was positive increased in the area production and yield under pulses. Among the seven districts of Meghalaya, East Garo Hill district registered the highest positive growth in area and production of pulses while east Khasi hill district registered highest growth rate in yield of pulses. The decomposition analysis indicated that interaction effect has the highest contribution to the growth in the production of pulses. Highest contribution was found to be area effect in West Garo hills district followed by East Garo Hill and Jaintia Hill district of Meghalaya. While yield effect was more in Ri-Bhoi district followed by Jaintia Hill and East Khasi Hill district. The contribution of interaction (area and yield) effect was more in east Khasi hill followed by West Khasi Hills district. Therefore, study suggested that attention is required in providing technical assistance in development of hybrid varieties, scientific management practice which will lead to increase in the pulses area and productivity. Government initiatives are required to give financial support and marketing facility for pulses. Hence, it is concluded that the pulse production can play a good role not only in the issue of food and nutritional security in Meghalaya but also will be helpful in reduce poverty, hunger, malnutrition and generate the income and employment to the farmer if proper policy initiatives are taken.

REFERENCES

Dandekar, V.M. (1980). Introduction to the seminar on data base and methodology for the study of growth rates in agriculture. *Indian J. of Agric. Econ.*, 35(2): 1-12.

FAO. (2017). Food and Agricultural Organization of United Nations. Economic and Social department of the Statistical division.

GoI. (2015). Commodity profile for pulse - February 2015. Department of Agriculture and Cooperation, Ministry of Agriculture, Government of India.

GoI. (2017). Agricultural Statistics at a Glance. Ministry of Agriculture and Farmer's Welfare, Department of Agriculture, Co-operation and Farmer's Welfare, Directorate of Economics and Statistics, Krishi Bhawan, New Delhi, India.

GoM. (2016). Directorate of Horticulture, Shillong Meghalaya

Joshi, P.K., and Rao, P.P. (2017). Global pulse scenario: status and outlook. *Annals of the New York academy of sciences.* 6(17): 6-17.

Kumar, B., Sharma, R., Singh, S.B., Shukla, L. and Khare, T.R. (2014).Effect of post-emergence application of Cyhalofop-butyl for weed management in direct seeded rice (Oryza sativa).*Indian Journal of Agricultural Sciences*, 84(8): 1018-1021.

Meena, N.K., Singh, R. Singh, N.U., Kumar, R. and Gurjar, G.N. (2017). Availability and requirements of major food items in NEH regions. *International journal of microbiology research,* 9(3): 864-868.

Palanisami,K.,Paramasivam,P. andRanganathan, C. R. (2002). Agricultural production economics: Analytical methods and applications, Associated Publishing Company, New Delhi.

Roy, A., Dkhar, D.S., Tripathi, A.K., Singh, N.U., Kumar, D. Das, S.K., Debnath, A. (2014). Growth performance of agriculture and allied sectors in North-East India. Economic Affairs. 59: 783795.

Roy, A., Singh, N.U., Dkhar, D.S., Mohanty, A.K., Singh, S.B., and Tripathi, A.K. (2015). Food security in North-East Region of India-A state wise analysis. *Agriculture Economics Research Review*. 28: 259-266.

Sharma, K. L. (1977). Measurement of the effects of area, yield and prices in the increases of value of crop output in India. *Agric. Situation India*, 32(6): 349-351.

Chapter 6

Nutrient Requirement of Papaya (*Carica papaya* L.) in the Homestead Farming System of Kerala for Yield Optimisation

B. Bindu

Farming Systems Research Station, Sadanandapuram, Kerala Agricultural University, Kerala

Papaya has gained commercial importance over the years because of its varied uses, mainly for table purpose. It is usually grown as homestead crop in Kerala. In recent years, isolated attempts have been made by some progressive farmers for commercial cultivation. Major production constraint encountered in papaya is difficulty in maximizing yield with in unit time. Balanced nutrition plays a vital role on plant growth, yield and fruit quality. One of the reasons for low production in papaya is inadequate nourishment. As the export of papaya from India is rapidly increasing in the recent past, there is a pressing need to enhance its productivity and improve the fruit quality. Papaya is a heavy feeder and it demands nutrients continuously in large amounts. Papaya variety CO-2 was used for the experimental purpose. Soil samples were analysed for available nitrogen, phosphous and potassium before and after the experiment. Nitrogen, phosphous and potassium content in leaf petiole were also analysed. The petiole samples were collected from the recently matured petiole. The trial was undertaken to study the effect of nitrogen, phosphorus and potassium application on growth, yield and quality of papaya and also to find out the optimum dose of NPK for commercial cultivation of papaya under Kerala conditions. The trial was conducted in 3^3 confounded factorial RBD, confounding NPK in replication-1 and NP^2K^2 in replication-2. Three different levels of nitrogen [200 (n_0), 250 (n_1), 300 (n_2) gram per plant per year], phosphorus [200 (p_0), 250 (p_1), 300 (p_2) gram per plant per year] and potassium [300 (k_0), 400 (k_1), 500 (k_2) gram per plant per year] were applied to papaya plants in six equal split doses at two months interval. Two month old seedlings were used for transplanting. Fertilizer application started thirty days after transplantation of seedlings to the main field. Urea, Rock phosphate and Muriate of potash were used as sources of Nitrogen, Phosphorus and Potassium. The *treatments involved 27 different combinations of Nitrogen, Phosphorus and Potassium at different levels, their interactions and control. Biometric characters like height of plants, girth of plants, number of leaves, time of first flowering, time of harvest were noted. Yield characters like number of fruits per plant, fruit weight, fruit length and girth, fruit volume, pulp percentage, total yield per plant and papain yield were recorded. Results revealed that application of nitrogen, phosphorus and potassium at the rate of 250:250:500 gram per plant per year in six equal splits, at two months interval was economically viable and improved the growth, yield and quality of papaya*

Keywords: *Papaya, Nitrogen, Phosphorus, Potassium, Yield.*

Papaya (*Carica papaya* L.) has gained commercial importance over the years because of its varied uses. In Kerala it is grown in an area of 16640 ha with an annual production of 103420 tonnes and with average productivity of 6.2 t ha-1 (FIB, 2016). Nutritive value, high yielding potential, year round fruiting behavior and short pre-bearing period make papaya unique among fruit crops. It is a rich source of Vitamin A (2020 IU and moderate source of vitamin C (40 mg $100g^{-1}$). Fruits of papaya are recognized for papain, a proteolytic enzyme extensively used in cosmetics and pharmaceutical industries. Papaya is an ideal fruit for processing due to year round production and availability of fruits at cheaper rates. It is used for the preparations of jam, jelly, nectar, ice cream flavor, tooty- fruity and crystallized fruits. Due to awareness of multifold uses of papaya for table, processing and papaya extraction purposes, it is slowly emerging from the status of a homestead crop to that of commercial crop. The major production constraint encountered in papaya is difficulty in maximizing yield with in unit time. Balanced nutrition plays a vital role on plant growth, yield and fruit quality. Papaya is very responsive to the application of inorganic fertilizes along with organic manures. One of the reasons for low production in papaya is inadequate nourishment. Understanding the interrelationships among vegetative growth, yield and nutrient uptake will help to exploit the high yielding potential of papaya plants. As the export of papaya from India is rapidly increasing in the recent past, there is a pressing need to enhance its productivity and improve the fruit quality. However under Kerala conditions, no systematic attempts have been made on the requirement of nutrition of papaya. The experiment was carried out to study the response of balanced nutrition on yield and yield attributes of papaya and also to find out the optimum dose of NPK for commercial cultivation of papaya under Kerala conditions.

Materials and Methods

The study was carried out at College of Agriculture, Vellayani, and using papaya variety CO-2. The experiment was conducted in 3^3confounded factorial Randomized Block Design; confounding NPK in replication-1 and NP^2K^2 in replication-2. Three different levels of nitrogen [200 (n_0), 250 (n_1), 300 (n_2) gram per plant per year], phosphorus [200 (p_0), 250 (p_1), 300 (p_2) gram per plant per year] and potassium [300 (k_0), 400 (k_1), 500 (k_2) gram per plant per year] were applied to papaya plants in six equal split doses at two months interval. Two month old seedlings were used for transplanting. Fertilizer application started thirty days after transplantation of seedlings to the main field. Urea, Rock phosphate and

Muriate of potash were used as sources of Nitrogen, Phosphorus and Potassium. The treatments involved 27 different combinations of Nitrogen, Phosphorus and Potassium at different levels, their interactions and control. Biometric characters like height of plants, girth of plants, number of leaves, time of first flowering, and time of harvest were noted. Yield characters like number of fruits per plant, fruit weight, fruit length and girth, fruit volume, pulp percentage, total yield per plant and papain yield were recorded. Fruit quality characters were also recorded during the study. Benefit:Cost ratio was worked out. Shelf-life of fruits were noted. Soil samples from the experimental area were analysed before and after experiment for available nitrogen, phosphorus and potassium Nitrogen, Phosphorus and Potassium content of leaf petioles were also assessed the tissue samples were collected from recently matured petiole.

Results and Discussion

NPK interaction had significant influence on plant height at all stages of growth. Lower dose of nitrogen (200g plant^{-1}year^{-1}) combined with highest dose of phosphorus (300g plant^{-1}year^{-1}) and potassium (500g plant^{-1}year^{-1}) gave maximum plant height in papaya (Table 6.1). The result of the present study is in conformity with the observations of Suresh *et al.* (2010) and Auxilia *et al.* (2008) who observed

Table 6.1. Influence of Nitrogen, Phosphorus and Potassium on Biometric Characters of Papaya

Treatments	*Plant Height (6MAP) (cm)*	*Plant Height (12MAP) (cm)*	*Leaf Number (6MAP)*	*Leaf Number (12MAP)*	*Time for First Flowering (Days)*	*Time for Harvest (Days)*
T_1 ($n_0p_0k_0$)	82.2	236.9	17.4	30.3	208.9	239.5
T_2 ($n_0p_0k_1$)	86.0	242.8	14.0	25.9	172.9	226.7
T_3 ($n_0p_0k_2$)	163.9	267.8	23.5	37.6	204.6	252.1
T_4 ($n_0p_1k_0$)	93.2	227.1	12.6	22.1	151.2	254.6
T_5 ($n_0p_1k_1$)	105.6	207.5	17.9	28.9	137.5	231.8
T_6 ($n_0p_1k_2$)	161.6	266.8	19.3	33.3	208.4	260.3
T_7 ($n_0p_2k_0$)	153.4	231.6	12.2	27.4	163.3	253.7
T_8 ($n_0p_2k_1$)	105.5	240.1	15.9	29.3	194.4	241.3
T_9 ($n_0p_2k_2$)	198.8	291.1	19.8	32.6	154.8	246.4
T_{10} ($n_1p_0k_0$)	145.3	262.6	13.2	27.2	147.3	252.6
T_{11} ($n_1p_0k_1$)	129.8	210.6	19.2	31.6	185.7	257.6
T_{12} ($n_1p_0k_2$)	180.5	273.9	20.8	28.1	127.6	231.5
T_{13} ($n_1p_1k_0$)	190.9	265.7	15.4	24.8	197.1	259.6
T_{14} ($n_1p_1k_1$)	132.8	235.4	17.9	30.5	146.7	234.9
T_{15} ($n_1p_1k_2$)	147.6	280.6	14.9	34.1	119.5	211.8
T_{16} ($n_1p_2k_0$)	122.4	248.8	22.2	30.7	169.0	250.1

Treatments	*Plant Height (6MAP) (cm)*	*Plant Height (12MAP) (cm)*	*Leaf Number (6MAP)*	*Leaf Number (12MAP)*	*Time for First Flowering (Days)*	*Time for Harvest (Days)*
T_{17} ($n_1p_2k_1$)	163.6	279.5	14.9	26.4	153.9	233.3
T_{18} ($n_1p_2k_2$)	175.2	197.1	21.4	35.7	121.9	224.0
T_{19} ($n_2p_0k_0$)	129.9	250.6	15.6	37.5	219.6	253.1
T_{20} ($n_2p_0k_1$)	106.2	254.2	17.4	30.5	163.2	233.7
T_{21} ($n_2p_0k_2$)	139.4	246.7	24.4	37.1	141.3	241.9
T_{22} ($n_2p_1k_0$)	150.4	282.4	16.1	28.5	212.8	249.0
T_{23} ($n_2p_1k_1$)	120.3	230.4	12.5	24.7	172.6	256.7
T_{24} ($n_2p_1k_2$)	162.9	261.4	25.6	40.0	149.5	239.4
T_{25} ($n_2p_2k_0$)	153.7	268.2	20.4	28.7	139.7	222.2
T_{26} ($n_2p_2k_1$)	134.1	258.4	17.2	31.8	216.7	241.2
T_{27} ($n_2p_2k_2$)	133.1	262.8	19.5	38.5	153.8	250.9
T_{28}Control	69.8	163.2	11.2	21.7	224.3	266.5
SE	3.2	3.5	0.3	0.6	2.7	1.8
CD (0.05)	6.6	7.3	0.7	1.3	5.7	3.7
Treatment Vs control	S	S	S	S	S	S

S: Significant; MAP: Months after planting.

that in papaya, lower dose of nitrogen combined with higher dose of phosphorus and potassium showed synergistic effect, thus resulting in increased height of plants. The study also revealed that NPK interaction had significant influence on plant girth at all stages of growth. Probable reasons for better girth of papaya plant may be higher requirement of nitrogen, phosphorus and potassium and synergistic effect between nitrogen and phosphorus as observed by Babu(2003). Bertuzzi and Rodreguez (2006) also obtained similar result in papaya. Potassium probably stimulated the efficiency of nitrogen utilization in respect of growth. Also it was seen that highest level of nitrogen (300g plant^{-1}year^{-1}) and potassium (500g plant^{-1} year^{-1}) combined with medium dose of phosphorus (250g plant^{-1} year^{-1}) resulted in maximum number of leaves. Lowest duration of flowering and harvesting was observed with NPK combination of 250:250:500g plant^{-1}year^{-1}. While control plants registered maximum number of days for harvest.

Nitrogen at 250 gplant^{-1}, phosphorus at 250 g plant^{-1}and potassium at 500g plant^{-1}had increased the number of fruits plant^{-1} by way of mean effect as well as interaction effect of nutrients (Table 6.2). This result was in conformity with the findings of Cruz *et al.* (2004) who observed that in papaya application of 500 g potassium gave significantly more number of fruitsplant^{-1}. Similar result was reported by Garcia *et al.* (2003) that in papaya, maximum number of fruits were obtained by applying nitrogen 200g, phosphorus 300g and potassium 500g per plant.

Table 6.2. Influence of Nitrogen, Phosphorus and Potassium on Yield Characters of papaya

Treatments	*Number of Fruits per Plant*	*Total Fruit Yield per Plant (kg)*	*Papain Yield (Kg/ha)*	*Fruit Weight (g)*	*Fruit Length (cm)*	*Fruit Girth (cm)*	*Benefit : Cost Ratio*
T_1 ($n_0p_0k_0$)	29.3	32.3	356.8	1163.7	25.9	40.6	2.1
T_2 ($n_0p_0k_1$)	21.3	19.8	415.9	959.2	21.9	38.1	1.8
T_3 ($n_0p_0k_2$)	16.8	22.5	397.6	1207.3	26.7	42.9	1.4
T_4 ($n_0p_1k_0$)	30.6	35.6	251.9	835.1	24.9	37.0	2.4
T_5 ($n_0p_1k_1$)	20.9	17.4	264.9	855.9	25.9	34.4	1.3
T_6 ($n_0p_1k_2$)	26.2	28.2	392.5	999.2	25.4	42.5	1.8
T_7 ($n_0p_2k_0$)	21.3	27.1	300.6	866.8	25.3	26.4	1.7
T_8 ($n_0p_2k_1$)	28.05	17.8	323.6	708.3	20.2	34.1	1.1
T_9 ($n_0p_2k_2$)	17.7	23.5	266.2	920.5	27.9	42.1	1.4
T_{10} ($n_1p_0k_0$)	24.1	24.5	318.9	895.8	21.6	24.3	1.6
T_{11} ($n_1p_0k_1$)	15.9	37.1	314.3	1246.7	25.3	41.5	2.3
T_{12} ($n_1p_0k_2$)	35.0	27.7	605.6	899.5	23.7	36.9	1.8
T_{13} ($n_1p_1k_0$)	26.1	19.2	323.5	1116.9	22.7	33.8	1.2
T_{14} ($n_1p_1k_1$)	19.5	26.4	377.8	1018.8	24.9	36.4	1.6
T_{15} ($n_1p_1k_2$)	43.6	59.8	674.6	1338.7	32.9	42.3	3.6
T_{16} ($n_1p_2k_0$)	36.0	28.8	541.7	453.7	22.4	27.6	1.7
T_{17} ($n_1p_2k_1$)	24.5	31.9	420.7	1238.2	25.1	26.1	1.9
T_{18} ($n_1p_2k_2$)	32.2	20.6	569.0	927.5	23.4	29.4	1.2
T_{19} ($n_2p_0k_0$)	20.0	32.6	390.10	1190.8	22.8	35.3	2.0
T_{20} ($n_2p_0k_1$)	15.6	22.8	258.9	1152.5	22.4	38.9	1.4
T_{21} ($n_2p_0k_2$)	28.7	27.5	268.7	719.5	20.9	32.8	1.6
T_{22} ($n_2p_1k_0$)	24.2	21.8	346.2	936.2	23.9	35.4	1.3
T_{23} ($n_2p_1k_1$)	29.5	39.1	272.9	854.7	24.3	27.4	2.3
T_{24} ($n_2p_1k_2$)	24.7	36.2	339.9	849.1	23.6	35.3	2.2
T_{25} ($n_2p_2k_0$)	30.8	27.7	402.4	911.6	27.6	30.9	1.7
T_{26} ($n_2p_2k_1$)	34.5	35.4	276.4	658.2	22.1	29.4	2.1
T_{27} ($n_2p_2k_2$)	26.7	27.7	347.3	1171.4	26.6	42.1	1.6
T_{28} Control	11.2	5.9	58.7	395.6	16.1	23.5	0.5
SE	1.4	1.7	21.7	23.1	2.	2.71	
CD (0.05)	2.9	3.4	45.1	47.9	NS	5.6	
Treatment Vs control	S	S	S	S	S	S	

S: Significant.

Medium dose of nitrogen and phosphorus (250 gplant^{-1}) combined with highest dose of phosphorus (500 g plant^{-1}) gave maximum fruit weight. This may be due to better combinations of N, P and K. Ghosh and Taria (2007) reported that in papaya, nitrogen application increased the weight of each marketable fruit. Rajkumar *et al.* (2005) also noted that application of 250 g nitrogen, 110 g phosphorus and 415 g potassium per plant per year resulted in increased fruit weight. Fruit length was not significantly affected by various treatments.

Maximum fruit girth was obtained from 200 g N, 200 g P and 500 g K. Biswas *et al.* (1989) noted that in papaya variety Ranchi, application of potassium increased breadth of fruit. Highest pulp percentage and fruit volume was obtained from 250 g N, 250 g P and 500 g K (T15). Medium dose of nitrogen and phosphorus (250 gplant^{-1}) along with highest dose of potassium (500g plant^{-1}) gave highest total fruit yield and papain yield per plant. The possible explanation for higher yield in treatment $n_1p_1k_2$ could be a favourable combination of NPK which provided better vigour to the plants. There is a close relationship between vigour of plant and yield (Cruz *et al*, 2007). Canesin and Correa (2006) reported that in papaya application of potash increased yield significantly. Highest benefit:cost ratio (3.55) was obtained from the combination of 250 g N, 250 g P and 500 g K plant^{-1} year$^{-1.}$

Nitrogen and phosphorus application had no significant influence on TSS content of fruits. This finding is in conformity with the observations of Akinyemi and Akanda (2008) and Kumar and Gho (2003) who observed that in papaya, TSS was not affected by different levels of nitrogen. But potassium application had significant influence on TSS content of fruits. Highest dose of potassium (500 g plant^{-1} year^{-1}) gave highest TSS content of fruits. Lowest acidity was noticed with the combined application of 200g nitrogen, 300g phosphorus and 500g potassium plant^{-1} year^{-1} (Table 6.3). Highest carotenoid content was reported by the application of lowest dose of nitrogen (200 g plant^{-1} year^{-1}) combined with highest dose of phosphorus (300 g plant^{-1} year^{-1}) and highest dose of potassium (500 g plant^{-1} year^{-1}). The results from the experiment showed that nitrogen and phosphorus application had no significant influence on ascorbic acid content. While application of highest dose of potassium (500 g plant^{-1} year^{-1}) resulted in highest ascorbic acid content in fruits. Highest total sugar and reducing sugar was obtained with combined application of 200 g N, 300 g P and 500 g K plant^{-1} year^{-1}.In the studies it was observed that highest shelf life of fruits was obtained from combined application of 300g N, 200g P and 300 g K plant^{-1} year^{-1}.

NPK interaction had significant influence on leaf petiole nitrogen content. Highest leaf petiole nitrogen content was recorded by treatment $n_1p_1k_{2\ i.e.}$, by the combined application of 250g nitrogen, 250g phosphorus and 500g potassiumplant^{-1} year^{-1}. Similar result was obtained by Kudada and Prasad (2006) in papaya, that concentrations of petiole nitrogen increased with nitrogen applications.

Highest leaf petiole phosphorus content was obtained by the combined application of 300g nitrogen, 250 g phosphorus and 500g potassiumplant^{-1} year^{-1}. Babu (2003) revealed that in papaya, phosphorus fertilization raised the petiole concentrations of P, N, Mn and Zn. But highest leaf petiole potassium content was

Table 6.3. Influence of Nitrogen, Phosphorus and Potassium on Fruit Quality Characters of Papaya

Treatments	*TSS (Per cent)*	*Acidity (Per cent)*	*Total Caro- tenoids (mg $100g^{-1}$)*	*Ascorbic Acid (mg $100g^{-1}$)*	*Total Sugars (Per cent)*	*Reducing Sugars (Per cent)*	*Non Reducing Sugars (Per cent)*	*Shelf Life (Days)*
T_1 ($n_0p_0k_0$)	11.6	0.1	2.2	41.9	12.3	10.4	1.9	5.2
T_2 ($n_0p_0k_1$)	10.6	0.3	2.6	38.9	10.6	8.6	2.0	6.9
T_3 ($n_0p_0k_2$)	13.8	0.2	1.5	42.8	14.6	11.7	2.8	4.1
T_4 ($n_0p_1k_0$)	10.8	0.2	2.1	40.4	8.7	7.2	1.5	5.4
T_5 ($n_0p_1k_1$)	10.8	0.2	2.3	40.6	9.7	7.3	2.4	6.1
T_6 ($n_0p_1k_2$)	14.4	0.2	2.54	42.3	14.3	12.7	1.6	6.3
T_7 ($n_0p_2k_0$)	13.9	0.3	1.3	43.8	10.9	8.8	2.0	4.1
T_8 ($n_0p_2k_1$)	12.4	0.1	2.3	39.4	11.4	9.5	1.9	5.3
T_9 ($n_0p_2k_2$)	16.7	0.1	3.7	51.7	15.4	13.9	1.6	7.2
T_{10} ($n_1p_0k_0$)	14.8	0.2	1.8	47.6	12.7	9.7	2.9	4.0
T_{11} ($n_1p_0k_1$)	11.7	0.3	2.9	41.8	8.2	5.1	3.2	6.8
T_{12} ($n_1p_0k_2$)	12.0	0.4	1.6	44.5	13.0	12.1	1.6	5.6
T_{13} ($n_1p_1k_0$)	12.7	0.2	1.8	36.7	12.1	10.9	1.3	4.7
T_{14} ($n_1p_1k_1$)	10.8	0.2	1.6	43.4	7.7	6.4	1.3	3.4
T_{15} ($n_1p_1k_2$)	14.9	0.3	3.1	42.9	11.0	9.2	1.8	7.0
T_{16} ($n_1p_2k_0$)	12.6	0.3	1.6	44.9	8.5	6.5	2.0	3.7
T_{17} ($n_1p_2k_1$)	12.4	0.2	2.0	39.2	11.2	9.5	1.9	6.5
T_{18} ($n_1p_2k_2$)	10.0	0.3	2.4	48.9	12.8	10.1	2.8	4.1
T_{19} ($n_2p_0k_0$)	12.2	0.3	2.4	48.9	8.3	7.0	1.3	7.3
T_{20} ($n_2p_0k_1$)	10.0	0.2	2.3	44.6	11.2	9.1	2.2	4.0
T_{21} ($n_2p_0k_2$)	15.2	0.2	2.6	41.3	12.7	10.2	2.5	5.2
T_{22} ($n_2p_1k_0$)	11.4	0.1	2.2	41.2	10.5	8.3	2.1	6.7
T_{23} ($n_2p_1k_1$)	12.5	0.3	2.6	47.4	9.9	7.5	2.3	3.9
T_{24} ($n_2p_1k_2$)	11.5	0.1	1.6	41.2	13.3	12.1	1.3	5.7
T_{25} ($n_2p_2k_0$)	12.3	0.15	2.5	45.9	9.6	8.4	1.2	4.2
T_{26} ($n_2p_2k_1$)	9.9	0.2	2.1	39.3	8.6	6.4	2.3	6.3
T_{27} ($n_2p_2k_2$)	13.0	0.2	1.9	47.2	13.5	11.7	1.8	4.9
T_{28} Control	7.5	0.4	0.8	31.8	5.6	4.4	1.3	3.2
SE	1.0	0.1	0.3	2.5	0.2	0.4	0.3	0.1
CD (0.05)	NS	0.08	0.7	NS	0.5	0.2	NS	0.2
Treatment Vs control	S	S	S	S	S	S	S	S

S: Significant.

obtained by the combined application of 200g nitrogen, 300 g phosphorus and 500g potassiumplant^{-1} plant^{-1}

Conclusion

The experiment revealed that application of nitrogen, phosphorus and potassium increased plant height, girth and number of leaves. Plants receiving a dose of nitrogen at 250 g, 300 g phosphorus and 500g potassium per plant took the shortest time for flowering. Combined application of nitrogen at 250 g plant^{-1}, phosphorus at 250 g plant^{-1} and potassium at 500g plant^{-1} considerably shortened the time for harvesting the first fruit. Fruit weight, number of fruits per plant, yield per plant and papain yield increased by the application of 250 g N, 250 g P and 500g K plant^{-1}year^{-1}. Levels of nitrogen and phosphorus had no significant influence on TSS and ascorbic acid content of fruits. Nitrogen at 200g, phosphorus at 250 g and potassium at 500g plant^{-1}was found to increase the shelf life of fruits. Application of nitrogen, phosphorus and potassium increased the soil and leaf petiole content of respective elements. Highest benefit: cost ratio was obtained from the combination of 250 g N, 250 g P and 500g K plant^{-1}year^{-1}.The overall assessment of the effect of major plant nutrients on papaya indicated that the application of nitrogen, phosphorus and potassium at the rate of 250:250:500 g plant^{-1} year^{-1} in six equal splits at two months interval was economically viable and improved growth, yield and quality of papaya under Kerala conditions.

REFERENCES

Akinyemi S and Akanda M O (2008). Effect of organic and inorganic fertilizers on growth and yield of pawpaw (*Carica papaya* L.). *Biological Agric Hort* 23(4): 383-392

Auxilia J, Balamohan T N and Nalina L (2008). Standardisation of stage wise nutrient requirement of papaya. In: *Proceedings of Second International symposium on papaya*, 9-12 December 2008, TNAU, Coimbatore, India, pp.99

Babu N (2003). Effect of organic inputs and potassium on growth and yield of CO-5 papaya (*Carica papaya* L.). *Indian J of Agric Sci* 73(4): 212-214

Bertuzzi S M and Rodreguez V A (2006). Response of pawpaw to application of nitrogen, phosphorus and potassium on a sandy soil in north western Corrientes. *Horticultura Argentina*15(2): 65-66

Biswas B, Sen S K and Maity S C(1989). Effect of different levels of nitrogen and potassium on growth, yield and quality of papaya variety Ranchi. *Haryana JHort Sci* 18(4): 17-20

Canesin R C and Correa F (2006). The use of manure associated with mineral fertilization of papaya seedling production. *Rev Bras Frutic*28(3): 483-486

Cruz J, Coelho E F and Santos M T (2004). Growth, dry matter and carbon partitioning of papaya in response to nitrogen nutrition. *Bragantia* 63(3): 351-361

Cruz J L, Pelcani C R and Carvalho J E (2007). Nitogen levels and photosynthetic rate of papaya cv Golden. *Ciencia Rural*37(1): 67-71

FIB (2016). Farm Guide 2016. Farm Information Bureau, 15th edn. Government of Kerala, Thiruvananthapuram.

Garcia E J L, Saucedo V C, Sanchez G and Soto H (2003). Mineral and foliar fertilization on development and production of papaya cv Maradol. *Revista Ciencia Agronomica*38(2): 136-141

Ghosh S N and Taria R K (2007). Effect of NPK on yield and quality of papaya in laterite soils. *Indian J of Fertil* 3(5): 47-49

Kudada N and Prasad S M (2006). Effect of planting season and manuring on incidence of papaya ring spot disease and fruit yield. *Indian JVirol* 17(1): 47-49

Kumar K and Goh K M (2003). Nitrogen release from crop residues and organic amendments as affected by biochemical composition. *Commun Soil Sci Pl Anal* 34(5): 521-536

Rajkumar M, Karuppaiah P and Kandaswamy R (2005). Effect of calcium and gibberellic acid on post harvest behaviour of papaya cv CO-2. *Indian J Hort* 62(4): 327-331

Suresh C P, Nath S, Poduval M and Sen S K (2010). Studies on the efficacy of phosphate solubilizing microbes and VAM fungi with graded levels of phosphorus on growth, yield and nutrient uptake of papaya (*Carica papaya* L.). *Acta Hort* 851: 401-406

Chapter 7

Response of Plant Nutrients on Productivity, Economics and Nutrient Uptake of Maize (*Zea mays* L.)–Chickpea (*Cicer arietinum*) Cropping System

G.J. Patel[1], G.N. Thorat[2] and A.Y. Makani[3]

[1]Research Scientist, [2]Assistant Professor, [3]Research Associate, Tribal Research cum Training Centre, AAU, Devgadh Baria, Assam

Multiplicity of cropping system has been one of the main features of Indian agricultural and it is attributed to rainfed agriculture and prevailing socio-economic situations of farming community. It has been estimated that more than 250 double cropping system are followed throughout the country and based on rationale of spread of crops in each district in the country, 30 important cropping system have been identified. By considering above the fact, the field experiments were carried out during the rainy (kharif) and winter (rabi) seasons of 2019-20 on farmers' fields in Dahod and Dhanpur taluka of Dahod district on the basis of productivity situated in Gujarat Plains and Hills, using maize-chickpea cropping pattern to see that effect of crop response to plant nutrients on soil properties and crop yields. Randomized Block Design (RBD) was used for statistical comparisons. The experiment was conducted with seven different treatments of NPK (T_1: Control, T_2: N (100 kg/ha), T_3:NP (100:50 kg/ha), T_4: NK (100:50 kg/ha), T_5: NPK (100:50:50 kg/ha), T_6: NPK + $ZnSO_4$ (100:50:50 + 20 kg/ ha) along with farmer's practices (T7) of NP (95:40 kg/ha) in kharif maize followed by rabi chickpea cultivation.T_1: Control, T_2: N (20 kg/ha), T_3:NP (20:40 kg/ha), T_4: NK (20:20 kg/ha), T_5: NPK (20:40:20 kg/ha), T_6: NPK + $ZnSO_4$ (20:40:20 + 20 kg/ha) along with farmer's practices (T_7) of NP (20:30 kg/ha). At each site/village, 4 farmers were selected thus making 6 replications. Maize cultivar 'GM 6' and chickpea cultivar 'JG 14' were sown. Result revealed that significantly the highest System Equivalent Yield (SEY) was obtained from treatment T_6 i.e. Recommended dose of NPK (100:50:50 *kg/ha) + 20 kg $ZnSO_4$ (4855 kg/ha) with net return of Rs. 39358. The nutrient response was 8.40 kg grain per kg nutrient was achieved from the treatment T_6 (NPK + $ZnSO_4$ - 100:50:50 + 20 kg/ha) in Maize and (NPK- 20:40:20 + $ZnSO_4$ - 20 kg/ha) in chickpea. While, in case of NPK concentration, it was slightly increased in grain as well as straw in chickpea. Therefore, cropping patter of maize in kharif after chickpea in rabi season with recommended dose of fertilizer along with $ZnSO_4$ increase the production as well as nutrient.*

Keywords: *Productivity, Economics, Maize-chickpea cropping system, Nutrient uptake, Nutrient response.*

The agriculture based Indian Economy needs high agriculture income for the farmers who feed the 1.25 billion Indian populations. Not only this like developed countries who despite having much lesser area under cultivation, export number of farm products, India also needs to be a prominent exporter of agriculture produce. This will attract more and more people towards agriculture, the backbone of the country. But how to achieve the dream of substantially increasing the farm income is a big question. The Prime Minister of India, Shri Narendra Modi in his Independence Day speech on August 15, 2017 laid out the route map to double the agriculture income in next 5 years.

Development of the agriculture sector in India has focused primarily on raising agricultural output and improving food security. This strategy involved an increase in productivity through better technology and location specific crops and their improved varieties as well as use of quality seeds and efficient use of fertilizers.

Maize (*Zea mays* L.) is one of the most versatile emerging crops having wider adaptability under varied agro-climatic conditions. Globally, maize is known as queen of cereals because it has the highest genetic yield potential among the cereals. Maize is an important food and feed crop which ranks third after wheat and rice in the India and world. Because of its expanded use in the agro-industries, it is recognized as a leading commercial crop of great agro-economic value. Chickpea is also an important crop which suites to the rabi season in cropping sequence. Because of its cultivation after harvesting of nutrient-exhaustive maize crop under residual soil moisture, chickpea often suffers due to imbalanced crop nutrition. Maize (*Zea mays* L.) in *kharif* and chickpea (*Cicer arietinum* L.) in *rabi* is one of the important cropping sequences of India in both irrigated and rainfed areas (Ahlawat *et al.*2005). the same situation also seen in Dahod district of Gujarat.

Gujarat is the one of the important states of growing maize and chickpea crops in India. During 2018-19, area under *kharif* maize and *rabi* chickpea has been sown in around 111858 hectares and 28037 hectares respectively in Dahod district of Gujarat. (DAO, Dahod 2019-20). The present study under taken with the major objective is to how this cropping system responds to different doses of N P K application.

Materials and Methods

Data employed in this study were taken from on-farm experiments conducted with Kharif maize and rabi chickpea between 2019 - 2020 in the Dahod district of Gujarat under the jurisdiction of Anand Agricultural university of the All India

Coordinated Research Project (AICRP) on Integrated Farming Systems (IFS) On-Farm Research (OFR) by the Indian Council of Agricultural Research (ICAR)-Indian Institute of Farming Systems Research, Modipuram, UP, India.

The Dahod district is well-known as "Adivasi region" in Gujarat. The district lies between 22038′ to 23013′ North Latitude and 73059′ to 74008′ Longitude. This district's height is above 217 meter from sea level. Dahod district is bounded by Panchmahals District to the west, Vadodara district to the south, Jhabua district of Madhya Pradesh state to the east and Banswara district of Rajasthan state to the north.

The district has semi-arid climate with three distinguished seasons *i.e., kharif* (Mid June to October), winter (November to February) and summer (March to Mid June). The average rainfall varies 735 mm. The summer is very hot where in temperature ranges from 4° to 47°C for the couple of days. However, the average maximum and minimum annual temperature are 42.5 °C and 11.9 °C, respectively. In general, soils are coarse in texture, non-calcareous and neutral to mildy alkaline in reaction, alluvial in origin and Organic carbon, nitrogen and phosphorus content of the soil is medium, posses neutral pH and Electrical conductivity (EC) is low. Total geographical area and gross cropped area of the district are 371, 715 ha and 341,700 ha, respectively. The cropping intensity of the district is 159 per cent that appears to be less because large portion of cultivable area is under rainfed. The district has significant areas under other cropping systems of Maize-Maize, Maize-chickpea, Maize-wheat-green gram, Paddy-wheat, Soybean-wheat-green gram, Pigeon pea-summer ground nut, Castor, Pigeonpea.

Paddy, Wheat and other commercial crops like Cotton, Cereals, and Gram *etc.* are mainly grown by the farmers. Maize is the traditional food crop and maize-chickpea is identified as the predominant cropping systems in the district. The average productivity of maize and chickpea in the district are 1192 kg ha^{-1} and 799 kg ha^{-1}, respectively. The experiment was conducted in two blocks on the basis of productivity *i.e.,* Dahod (high productivity area) and Dhanpur (low productivity area) of Dahod district with six different locations taking four farmers from each location. The numbers of cultivator's fields selected for evaluation purpose in Dahod districts were twenty-four.

Block and Villages

Block	*Villages*
High Productive (Dahod)	Bawka, Bordi-Sarkari and Bhathiwada
Low Productive (Dhanpur)	Lukhadiya, Amli-menpur and Kanjeta

The farmer's field where maize-chickpea cropping system (MCCS) has been practiced for more than 10-12 consecutive years were selected to ensure that the research findings were relevant to long-term MCCS and not baffled by rotation and fertilizer management with other crops having different input demands. A stratified random sampling technique was used for selecting the experimental unit at every location. Farmer fields were considered as a replication at each location and

Randomized Block Design (RBD) was used for statistical comparisons. Descriptive statistical analysis was used for different parameters to establish the range of variability and deviation with in location using standard error of mean.

Selected farmer was interviewed between September and October of 2019 to examine their current crop and nutrients management practices. Seven fixed plots (100 m^2 each) were demarcated in every farmer's field to evaluate a set of seven fertilizer treatments both in maize and chickpea, the fertilizer treatments were Control (no fertilizer application), N alone (N-0-0-0), NP (N-P-0-0), NK (N-0-K-0), NPK (N-P-K-0), NPK and Zn (N-P-K-Zn), and Farmers fertilizer management practices (FFMP). As per the treatments the recommended dose of nitrogen (N), phosphorus (P), potassium (K), and zinc (Zn) was applied to both the crops under investigation except absolute control, where no application of fertilizers (nutrients) was sorted. The fertilizer treatments were assigned to maize crop randomly in each plot ensuring that the same level of fertilizer has come in subsequent chickpea crop of the system.

All fertilizer treatments were applied in the recommended dose except in Farmer's Field management practices (FFMP). The recommended dose of NPK and Zn were applied 100:50:50:20 kg/ha in maize and 20:40:20:20 kg/ha in chickpea. Fertilizers used as a source of nutrients included Urea, DAP, MoP and $ZnSO_4$.

Before the initiation of on-farm experiments during *kharif* 2019-2020 at the locations, soil samples were collected from the plough layer (0–15 cm depth) at ten places in each of the 24 farmer's fields. Collected soil samples were mixed thoroughly and well-homogenized by using a pestle and mortar and passed through a 100-mm sieve to eliminate the undesired material, afterwards the representative sample was drawn for chemical analysis. Soils were analyzed for organic carbon (Walkley and Black method), extractable N (alkaline $KMNO_4$ method) suggested by Subbiah, extractable P (0.5 M $NaHCO_3$, pH 8.5 extraction), and extractable K (1 M NH_4OAc, pH 7.0 extraction).

Results and Discussion

Productivity of Maize and Chickpea

Application of recommended doses of NPK along with $ZnSO_4$ to both the crops resulted in significantly higher grain and fodder/stover yields of maize and chickpea on (Table 7.1). The increase in grain yield of maize owing to application of recommended doses of NPK along with $ZnSO_4$ was 60,44,33,26,10 and 29 per cent higher over the control, N, NP, NK, NPK and farmers' practice respectively (Table 7.1) in case of fodder maize yield owing to application of recommended doses of NPK along with $ZnSO_4$ was 19, 36, 26, 18, 10 and 24 per cent. Further in succeeding chickpea, significantly higher seed and stover yields of chickpea were observed with application of NPK+$ZnSO_4$. The increase in grain yield with this treatment were to the tune of 40, 26,15,17,9 and 21 per cent higher over the control, N, NP, NK, NPK and farmers' practice respectively. in case of chickpea fodder yield owing to application of recommended doses of NPK along with $ZnSO_4$ was 29,18,6,9,10 and 12 per cent. Again, with respect to system productivity,

NPK fertilization along with recommended ZnSO4 to component crops recorded significantly higher maize-equivalent yield (4855 kg/ha) on account of higher grain/seed yield of component crops. Significant improvement in the grain/seed yield may be attributed to significantly higher uptake of nutrients (NPK) by both crops (Table 7.1) owing the optimum supply of plant nutrients. Further, soils of the experimental sites are deficient in Zn, the application of this deficit nutrient helped both the crops to record higher yields over NPK treatment alone (Table 7.2). This corroborates the findings of Sharma *et al.* (2012), Paramasivan *et al.* (2014) and Saha *et al.*(2015). The data also showed that application of N over the control, NP over N, NK over N, NPK over NP, and NK and NPK+$ZnSO_4$ over NPK had advantage in grain and fodder/stover yields of maize and chickpea. Among the nutrient combinations, NPK+$ZnSO_4$ application was found significantly superior to NP, NK, NPK and farmer's practices. The results are in agreement with the findings of Hile *et al.* (2007) and Jain *et al.* (2012).

Table 7.1. Influence of Treatments on Grain and Fodder Yield (kg/ha) Maize-Chickpea Cropping System

Treatments	*Kharif Maize*		*Rabi Chickpea*		*SEY*(kg/ha)*
	Grain Yield (kg/ha)	*Fodder Yield (kg/ha)*	*Grain Yield (kg/ha)*	*Fodder Yield (kg/ha)*	
T_1: Control	1269	1608	789	716	3298
T_2: N	1408	1756	873	783	3653
T_3: NP	1519	1903	957	870	3980
T_4: NK	1603	2033	943	850	4028
T_5: NPK	1842	2187	1012	915	4444
T_6: NPK + ZN	2021	2395	1102	995	4855
T_7: FP	1570	1939	910	825	3910
S. Em.+	33	38	17.50	17.30	-
CD (P=0.05)	92.39	107	49	48.45	-
C. V. (per cent)	10.7	9.49	9.11	9.97	-

Price of produce: Maize grain: Rs 14/kg, Maize fodder: Rs. 2.5/kg, Chickpea grain: Rs 36/kg, Chickpea straw: Rs. 2/kg

* Equivalent yield in terms of base crop, FP: Farmers Practice.

Fodder yield is not included in system equivalent yield.

MN: $ZnSO_4$

Application of 100 kg N, 50 kg P_2O_5, 50 kg K_2O and 20 kg $ZnSO_4$ to maize and application of 20 kg N, 40 kg P_2O_5, 20 kg K_2O and 20 kg $ZnSO_4$ to chickpea where, resulting significantly higher gross returns and system net returns in T_6: NPK + Zn over the remaining treatments including farmer's practices (Table 7.2). These findings are in line with those of Sharma *et al.* (2012) and Hiremath *et al.* (2016).

Table 7.2. Influence of Treatments on Nutrient Concentration in Grain and Straw under Maize–Chickpea Cropping System

***Kharif* Maize**

Treatments	*Grain*								*Fodder*							
	Nutrient Concentration (per cent)				*Sufficient (+)/Deficient (-)*				*Nutrient Concentration (per cent)*				*Sufficient (+)/Deficient (-)*			
	N	*P*	*K*	*MN**	*N*	*P*	*K*	*MN**	*N*	*P*	*K*	*MN**	*N*	*P*	*K*	*MN**
T_1: Control	1.39	0.28	0.49	-	+	+	+	-	0.56	0.13	1.17	-	+	+	+	-
T_2: N	1.43	0.30	0.49	-	+	+	+	-	0.60	0.14	1.18	-	+	+	+	-
T_3: NP	1.49	0.30	0.50	-	+	+	+	-	0.58	0.18	1.17	-	+	+	+	-
T_4: NK	1.48	0.31	0.51	-	+	+	+	-	0.60	0.16	1.33	-	+	+	+	-
T_5: NPK	1.50	0.34	0.54	-	+	+	+	-	0.60	0.18	1.40	-	+	+	+	-
T_6: NPK + MN	1.51	0.38	0.56	-	+	+	+	-	0.62	0.20	1.47	-	+	+	+	-
T_7: FP	1.50	0.33	0.52	-	+	+	+	-	0.60	0.17	1.27	-	+	+	+	-

*Concentration of micronutrients was not determined.

***Rabi* Chickpea**

Treatments	*Grain*								*Straw*							
	Nutrient Concentration (per cent)				*Sufficient (+)/Deficient (-)*				*Nutrient Concentration (per cent)*				*Sufficient (+)/Deficient (-)*			
	N	*P*	*K*	*MN**	*N*	*P*	*K*	*MN**	*N*	*P*	*K*	*MN**	*N*	*P*	*K*	*MN**
T_1: Control	3.10	0.61	0.71	-	+	+	+	-	1.09	0.19	1.14	-	+	+	+	-
T_2: N	3.29	0.63	0.76	-	+	+	+	-	1.13	0.21	1.19	-	+	+	+	-
T_3: NP	3.35	0.67	0.75	-	+	+	+	-	1.20	0.24	1.20	-	+	+	+	-
T_4: NK	3.34	0.64	0.81	-	+	+	+	-	1.17	0.22	1.24	-	+	+	+	-
T_5: NPK	3.29	0.68	0.85	-	+	+	+	-	1.20	0.26	1.26	-	+	+	+	-
T_6: NPK + MN	3.36	0.72	0.90	-	+	+	+	-	1.26	0.28	1.30	-	+	+	+	-
T_7: FP	3.28	0.66	0.82	-	+	+	+	-	1.13	0.23	1.14	-	+	+	+	-

*Concentration of micronutrients was not determined.

Nutrient Uptake by Maize and Chickpea

Nutrient uptake by any crop is the product of nutrient concentration in the tissue and the dry-matter production potential. In the present study, an application of NPK+$ZnSO_4$ at recommended rates to maize recorded significantly higher N, P and K uptake at harvesting, followed by recommended NPK (Table 7.3). Similarly, in chickpea, an application of recommended NPK along with $ZnSO_4$ recorded significantly higher uptake of N, P and K over rest of the treatments. The total uptake of nutrients (NPK) by maize–chickpea cropping system was the maximum under NPK+$ZnSO_4$ fertilization at recommended doses (Table 7.3). This could be attributed to the fact that added nutrients increased the N, P and K concentrations in grain and stover/haulm of both the crops by providing balanced nutritional environment inside the plant and higher photosynthetic efficiency which favored better growth and crop yield. Jain *et al.* (2012) also reported higher uptake of N and P by grain and stover of maize–wheat cropping system with the application of fertilizers.

Table 7.3. Influence of Treatments on Grain Yield (kg/ha), Economics (Rs/ha) and Nutrient Response (kg/kg and Rs/Rs) of Crops (Maize-Chickpea)

Treatments	*Gross Income Rs/ha*	*System Cost of Cultivation*	*System Net Returns*	*Nutrient Response*	
				kg grain/ kg nutrient	*Rs/Rs*
T_1: Control	51625	19793	31832	0.00	0.00
T_2: N	57101	23580	33521	9.35	3.58
T_3: NP	62227	29757	32471	7.98	2.15
T_4: NK	63168	24783	38384	7.29	3.92
T_5: NPK	69519	30575	38944	7.68	2.61
T_6: NPK + MN	75923	36565	39358	8.40	2.05
T_7: FP	61223	31089	30133	8.33	1.82

Price of produce: Maize grain: Rs 14/kg, Maize fodder: Rs. 2.5/kg, Chickpea grain: Rs 36/kg, Chickpea straw: Rs. 2/kg

* Equivalent yield in terms of base crop, FP: Farmers Practice.

Fodder yield is not included in system equivalent yield.

MN: $ZnSO_4$

Conclusion

Based on the results of experimentation revealed that significantly the highest System Equivalent Yield (SEY) was obtained from treatment T_6 *i.e.* Recommended dose of NPK (100:50:50 kg/ha) + 20 kg $ZnSO_4$ (4855 kg/ha) with net return of Rs. 39358. The nutrient response was 8.40 kg grain per kg nutrient was achieved from the treatment T_6 (NPK + $ZnSO_4$ - 100:50:50 + 20 kg/ha) in maize and (NPK- 20:40:20 + $ZnSO_4$ - 20 kg/ha) in chickpea. While, in case of NPK concentration, it was slightly increased in grain as well as straw in chickpea. Therefore, cropping patter of maize in *kharif* after chickpea in *rabi* season with recommended dose of fertilizer along with $ZnSO_4$ increase the production as well as nutrient.

REFERENCES

Hile, R.B., Patil, H.M., Patil, Y.J. and Bhosale, S.S. (2007). Effects of N, P and K on productivity and soil fertility in maize–wheat cropping system. *International Journal of Agricultural Science* 3(2): 205–207

Jain, N.K., Singh, Hari and Dashora, L.N. (2012). On-farm response of maize–wheat cropping system to fertilizer NPK input. *Research on Crops* 13(2): 475–480.

Paramasivan, M., Kumaresan, K.R. and Malarvizhi, P. and Velayudham, K. 2014. Effect of different levels of NPK and Zn on yield and nutrient uptake by hybrid maize in Pilamedu and Aplaviduthi series of Tamil Nadu. *Madras Agricultural Journal* 98(10–12): 334–338.

S. M. hiremath, R. Mohan kumar and Ashok Kumar Gaddi (2016). Influence of balanced nutrition on productivity, economics and nutrient uptake of hybrid maize (*Zea mays*)–chickpea (*Cicer arietinum*) cropping sequence under irrigated ecosystem. *Indian Journal of Agronomy* 61 (3): 292-296.

Saha, B., Saha, S., Hazra, G.C., Saha, S., Basak, N., Das, A. and Mandal, B. (2015). Impact of zinc application methods on zinc concentrations and zinc use efficiency of popularly grown rice (*Oryza sativa*) cultivars. *Indian Journal of Agronomy* 60(3): 391–402.

Sharma, S.K., Jain, N.K. and Upadhyay, B. (2011). Response of groundnut (*Arachis hypogaea* L.) to balanced fertilization under sub-humid Southern plain zone of Rajasthan. *Legume Research* 34: 273–277.

Chapter 8

Role of Krishi Vigyan Kendra for Transfer of Technology under Front Line Demonstrations in Tribal Area of Jalgaon District

M.V. Mahajan, S.J. Mahajan and D.S. Nehte*

Krishi Vigyan Kendra, Pal (Jalgaon-I), Maharashtra

The Krishi Vigyan Kendras (KVK) is of national importance which would help in accelerating the agricultural production and also in improving the socio-economic conditions of the farming community. Krishi Vigyan Kendra, Pal under Central Plateau Zone (assured rainfall). Need based training programmmes are being conducted by the KVK in various disciplines such as agronomy, horticulture, plant protection, women in agriculture, agricultural engineering and others. Survey method by personal interview was used for collecting the required information on component wise crop production technology transferred, extent of adoption of technology, Frontline demonstration, discipline wise training organized by KVK and farmers response towards training, costs and returns of crop production. It was concluded that KVK, Pal playing a vital role in disseminating the improved crop production technology and helps in increasing the crop yield. Further research need to be focused on the problems and constraint for adoption of the technology.

Keyword: *Krishi Vigyan Kendra, Training, Transfer of technology, Frontline demonstration, Adoption, Crop yield.*

Under the age is of the Indian Council of Agricultural Research (ICAR), New Delhi, a Krishi Vigyan Kendra was set up under the management of Satpuda Vikas

Mandal, Pal on August, 1st 1984. This NGO is chaired by Hon'ble Shri. Balasaheb Madhukarrao Chaudhari, Ex-Chairman of Legislative Assembly, of Maharashtra State. The Host Institute is working in remote tribal area since 1949 in the field of education, health, infrastructure development, agriculture and subsidiary occupation, cottage industries more particularly for tribal people.

Krishi Vigyan Kendra is an institutional project of the Indian Council of Agricultural Research (ICAR) to demonstrate Science and Technology input of Agricultural Research and education on the farmer's field and in the rural areas with the help of multi-disciplinary team of scientists. The KVK falls into the jurisdiction of MPKV, Rahuri. The climate of the district is almost dry except in the monsoon. The average rainfall of the district is about 700 mm. The temperature ranges from 6° to 10° C in winter to 42° to 47° C in summer. The district lies between 20° and 21° C north latitude and 75° 55′ and 76° 28′ east longitude with total area of 11,765 sq km.Need based training programmmes are being conducted by the KVK in various disciplines such as agronomy, horticulture, plant protection, soil science, women in agriculture, agricultural engineering and others. The present study focused on how KVK is transferring the technology through training and demonstration and response of farmers in adoption of technologies with economic viability.

Materials and Methods

The study was confined to the two blocks of the Jalgaon districts such as Raver, Yawal. In these blocks maximum numbers of farmers were benefited by the KVK. Five villages namely Garbardi, Grakheda, Nimdya, Jamnya and Gadarya from these two blocks and 100 farmers who were adopted by KVK, Pal for transfer of technology during Kharif and Rabi season were selected from these villages for detailed survey. Different components of crop production, horticulture, Plant protection and women in agriculture were given by the scientists of the KVK, Pal to the selected farmers. The required information was collected by survey method by personal interview on component wise of crop production technology. The required information were transfer of technology, extent of adoption of technology by contact farmers, discipline wise training organized by KVK and farmers response towards training, costs and returns of crop production technology *etc.* to accomplished the objectives of the study. The data collected for this study refer to Kharif and Rabi season during the year 2018-19

Results and Discussion

Training achievements the main mandate of Krishi Vigyan Kendra is to conduct the variety of trainings for the benefit of farmers, rural youth and extension personnel of the district. Training programme helps the farmers to aware the latest technical knowledge and skill related to agriculture and allied field. KVK training programme starts with identification of training needs in the villages. Then the training programme conducted to improve the knowledge and developed new skills required for adoption of the latest technology and build up scientific attitude among farming community. KVK, Pal conducted the training programme in five disciplines such as agronomy, horticulture, plant protection, agricultural engineering, Livestock management and women in agriculture (Table 8.1). Trainings on plant protection were very much useful because crop protection status of the blocks is poor. As the farming communities are generally tribal and resource poor, they are very much interested in small farm implements. They are also lacking the knowledge of the use of improved farm implements. Trainings on women in agriculture were also very much useful. This may be due to the fact that specialized and intensive practical training was imparted. Training on agronomy, horticulture and soil science were also useful but less as compared to the agriculture engineering and home science discipline.

Table 8.1. Achievements of Training Programme for Practicing Farmers and Farm Women of KVK, Pal (Jalgaon-I) during 2018-19

Sl. No.	*Particulars*	*Agronomy*	*Horti-culture*	*Plant Protection*	*Animal Husbandry and Dairy Science*	*Women in Agriculture*	*Total*
1	No. of trainings	11	12	12	09	08	
2	Duration in days	18	19	14	12	11	
3	No. of participants						
	i. General						
	Male	12	14	14	11	04	55
	Female	09	07	05	09	10	40
	ii. SC						
	Male	52	55	57	61	00	225
	Female	37	33	49	36	42	197
	iii. ST						
	Male	64	60	69	71	65	329
	Female	54	47	51	55	67	274
	iv. OBC						
	Male	48	72	84	73	114	391
	Female	55	68	79	63	57	322
	Total	331	356	408	379	359	1833

Technology Transfer and Adoption

The data on transfer of technology under KVK, Pal were collected in season wise under the different thematic area/components such as integrated crop management, integrated weed management, integrated nutrient management, fruit crop cultivation, vegetable crop cultivation, plant protection, women empowerment, livestock management and farm mechanization. The data collected from the selected respondents are presented in the Table 8.2. Some demonstrations were conducted on green gram, black gram, red gram, vegetables and fruit crop for better awareness of the improved cultivation practices during *Kharif* and *Rabi* season. On thematic area basis highest adoption were found in integrated pest and disease management and then Fruits crop cultivation and Livestock management. This implies that farmers are very much interested in use of plant protection techniques and Fruits crop cultivation in agriculture. It was also found that adoption were little higher during *Rabi* than *Kharif* season.

Table 8.2. Technology Transfer and Adoption by the Respondents in different Thematic Area-wise during Kharif and Rabi Season during 2018-19

Sl. No.	*Thematic Area*	*Kharif (n=100)*			*Rabi (n=100)*		
		C	*P*	*N*	*C*	*P*	*N*
1	Integrated crop management	33	48	19	34	50	16
2	Integrated weed management	31	52	17	32	53	15
3	Integrated nutrient management	37	47	16	33	47	20
4	Integrated pest management	56	31	13	58	32	10
5	Integrated disease management	52	39	09	54	38	08
6	Fruits crop cultivation	51	34	15	53	36	11
7	Vegetable crop cultivation	49	39	12	53	39	08
8	Post-harvest technology	33	54	13	37	49	14
9	Livestock management	49	34	17	39	42	19
10	Women empowerment	36	44	20	37	46	17
	Adoption index (per cent)	42.70	42.20	15.10	43.00	43.20	13.80

C: Complete; P: Partial; N: Nil.

Constraints for Poor Adoption

During training programme and field visit, some problems or constraint faced by the farmers for adoption of new technology has been studied. These constraints may be useful to the KVK for new technology and also refinement of the technology. The major constraints for the poor adoption of technology are marketing and storage facility of the farm produce. The unavailability of improved farm implements and their service provider is also the major problem for poor adoption of farm mechanization. All the major constraints are givenin the Table 8.3.

Table 8.3. Constraints for the Poor Adoption of the Technology

Sl.No.	*Constraints/Problems for Poor Adoption of Technology*	*Per cent Respondents Reported*
1.	Lack of marketing facility	83
2.	No storage facility	74
3.	Use of Tractor drawn implements for small farmers	81
4.	Unavailability of service provider for agro inputs	77
5.	Unavailability of improved farm implements	64
6.	Illiteracy of the farmers	64
7.	Water scarcity	53

Effect of Technology on Crop Production

Table 8.4 indicated the increased in production of the different crops after adoption of the improved package and practices under frontline demonstration in the KVK adopted villages. The increased in yield in HYV of Tur variety BDN-711 was 26.86 per cent as compared to local variety. The B:C ratio of the tur was found to be 6.97 The increase in yield of Ground nut was found to be 37.07 as compared to local variety and B:C ratio was 5.65. Increase in yield of 28.81 per cent was found to be in case of soybean production. Similarly increase in yield of 40.94 per cent was found to be in case of Moong cultivation. The yield increased by 2815.60 per cent and B:C ratio was found to be 6.91. In case of rabbi gram the increase in yield was found to be 30.66 per cent.

Table 8.4. Effect of Technology on Economics of FLD on Crop Production

Crop	*Variety*	*Economics of Demonstration (Rs./ha)*				*Yield qt/ha*		*Per cent Increase in Yield*
		Cost of Culti-vation (Rs./ha)	*Gross Return (Rs./ha)*	*Net Return (Rs./ha)*	*B:C*	*Demo*	*Check*	
Tur	BDN 711	10500	83750	73250	6.97	16.75	12.25	26.86
Moong	Utkarsha	7960	41148	33188	4.16	7.62	4.5	40.94
Udid	TAU-1	6880	54432	47552	6.91	7.56	5.2	15.6
Soyabean	JS-9560	12600	61600	49000	3.88	22	15.66	28.81
Groundnut	JL-776	19700	131100	111400	5.65	21.85	13.75	37.07
Bengal gram	Digvijay	16110	84880	68770	5.27	21.22	16.24	30.66

Conclusions

Hence, it can be concluded that KVK, Pal playing a vital role in disseminating the improved crop production technology and helps in increasing the crop yield. The technology transferred is also profitable and acceptable to the farming community. Further research need to be focused on the problems and constraint for adoption of the technology

REFERENCES

Behera, S.K., Maharana, J.R. and Acharya, P. (2014).Transfer of Technology through Krishi Vigyan Kendra for theTribal Farmers in Hilly Areas of Koraput District. *Indian Journal of Hill Farming*, 27(2): 34-37.

Nazir, T., Vaida, N. and Dar, M.A. (2013). Role of Krishi Vigyan Kenderas for the empowerment of rural women through vocational trainings. Basic Research *Journal of Agricultural Science and Review*, 2(3): 51-56.

Rathi, D. and Singh, S.P. (2010). Farmers' Perception about Krishi Vigyan Kendra in Satna District of Madhya Pradesh. *Journal of Rural and Agricultural Research*, 10(2): 56-59

Sharma, N., Arora, R.K. and Kher, S. (2010). KVK trainings for the farmers in hilly areas of Poonch district – identifying need of the hour. *Journal of Hill Agriculture*, 1(2): 140-145.

Sridhar, G., Malleswar, S.S.N., Rao, B.S., Patil, D.V. and Prasad, N. (2013).Transfer of Technology through Agricultural Scientists-Some Special Success Stories (from on Farm Testing, Frontline Demonstrations and Trainings) and IndegenousTechnological Knowledges (ITKs) Identified by KVK, Visakhapatnam District. *International Journal of Innovative Research and Development*, 2(10): 233-240.

Chapter 9

A Review Paper on Impact of Environmental Externalities in Agriculture and Allied Sector

***Rachana Kumari Bansal*[1]*, *Y.C. Zala*[2] *and A.S. Shaikh*[1]**

[1]Assistant Professor, Department of Agricultural Economics, B.A. College of Agriculture, Anand Agricultural University, Anand – 388 110, Gujarat
[2] Principal and Dean, International Agri-Business Management Institute (IABMI), Anand Agricultural University, Anand – 388 110, Gujarat

Externality prevents the attainment of Pareto-Optimality, whether it is positive or negative.In this review paper, positive and negative externalities and its impact on environment were documented as it differs across various production systems. The positive externalities associated with livestock production in India's mixed farming system includes saving of chemical fertilizers due to use of dung as manure and prevention of carbon dioxide emission (4.17 Mt) due to use of animal energy in agriculture. The study has found the saving of soil nutrients (0.54 Mt of N, 0.14 Mt of P and 0.54 Mt of K) and saving of chemical fertilizers (2.63 Mt of ammonium sulphate, 0.31 Mt of super phosphate and 0.82 Mt of Murate of potash) in Indian mixed farming system on account of use of dung as manure. Mangrove ecosystem and its impact was seen in the form of positive externalities like- shoreline protection, increased fish catch and enhanced income, utilization of non-timber forest products, control of soil erosion and water retention. Besides nitrogen fixation, the positive impact of pulses was seen in the form of reduction of farmers expenses and burden of government subsidy on fertilizers and environment because of reduced production and use of synthetic urea. Pulses reduced the expenditure on transportation and urea application in field and 10 per cent application losses, farmers could save Rs 613 ha and also reduced the burden of Government subsidy by Rs 1221/ha. Further, it also reduced the potential damage to environment by 1137 Rs/ha in terms of greenhouse gases and health effects. The total positive externality of pulse farming was estimated as Rs 2971/ha.

The negative externalities were seen in the form of effluents discharged by the industries as well as use of pesticides, had led to severe pollution, which had ultimately affected the livelihood of the poor. It can be seen in the form of human health hazards (occupational poisoning and pesticide residues through contamination) and reducing the value of land and agricultural production on affected farms. The value of a highly affected farm land was less than half of a farm land less affected by pollution. Losses of beneficial insects, loss of biodiversity and ecosystem as well as honey bee loss and pest resistance were also appeared as a part of negative externality in the environment.

Keywords: *Biodiversity, Effluents, Externality, Expenses, Livelihood and Subsidy.*

Any actions of one economic agent which affects another but is not reflected in market price, leads to a situation which is called an externality. Externality can be either positive (apiculture leading higher pollination in adjacent fields) or negative (pollution affecting quality of life), prevents the attainment of Pareto-Optimality (Bhat and Umesh, 2016). When social and private costs and social and private benefits diverge, perfect competition will not achieve Pareto-optimality. *In case of external economies, the private marginal cost of the firm will be higher than the social marginal cost.* It will result in under-production of the product because it will not produce its product to the extent social interest requires and loss of social welfare occurs. But, when external diseconomies occur, prices fixed on the basis of private marginal cost will be lower than that determined on the basis of social cost. It causes over production of the product and loss of social welfare.

Reviews of Literature

Livestock and its Impact on Environment

Impact of livestock on environment, however, differs across production systems. Industrial livestock production systems cause more harm to environment, while mixed crop livestock systems are benign to environment (Sere and Steinfield, 1996). Positive externalities of livestock to environment are less documented and non-quantified but the some prominent contributions are included here like.

Table 9.1. Quantity of Diesel Saved Due to Use of Animal Energy from Indian Mixed Farming System

Parameters	*Values*
Number of draught animals (millions)	55
Replacement rate	10.0
Number of tractors required to replace existing stock of working animals (millions)	5.463
Number of operating hours (hours/year)	801
Consumption of diesel by required no. of tractors (Mt)	13.13
Total carbon released from burning of estimated diesel use (Mt)	11.38
Emission of carbon dioxide prevented (Mt)	4.17
Emission prevention equivalent to methane (Mt)	0.198

Swamy and Bhattacharya (2006) using time series data, have arrived at a substitution rate of 10, that is a tractor could replace 10 working animals and at this rate, approximately 5.5 million tractors would be required to replace the existing stock of working animals *i.e.* 55 millions. To use the services of required number of tractor stock, approximately 13 Mt of diesel would be required annually. Burning of this much diesel would emit about 4.17 Mt of CO_2, or 0.199 Mt of methane, which is the methane emission prevented by the working animals.

Dixit and Birthal (2013) found that in India, the livestock helps in saving natural resources through their synergistic relationship with cropping activities. They also quantified the positive environmental externalities associated with livestock production in India's mixed farming systems through their study.

Table 9.2. Saving of Chemical Fertilizers Due to Use of Dung of Livestock as Manure in Mixed Farming Systems of India

Parameters	*Values*
Production of wet dung	635 Mt
Dung utilized as manure	381 Mt (60 per cent of total production)
Dung manure on dry matter basis	76.2 Mt (20 per cent of dung as manure)
Savings of soil nutrients (Mt)	
Nitrogen (N)	0.541
Phosphorus (P)	0.137
Potash (K)	0.541
Total (N, P and K)	1.219
Saving of chemical fertilizers (Mt)	
Ammonium sulphate	2.63
Super phosphate	0.31
Murate of potash	0.82

They found that there was saving of chemical fertilizers (2.63 Mt of ammonium sulphate, 0.31 Mt of super phosphate and 0.82 Mt of Murate of potash) in Indian mixed farming system on account of use of dung as manure. About 76 Mt of dung (on dry matter basis) was used as manure. The saving of soil nutrients had been estimated to the tune of 1.22 Mt comprising 0.54 Mt of N, 0.14 Mt of P and 0.54 Mt of K. The saving of these nutrients were apparently small, but its value in monetary terms could be substantial. Further, its environmental value can be gauged if we consider the associated emission in the production and transportation of the equivalent amount of chemical fertilizers at the farm gate.

Forest and its Impact

Mangrove forests, world's most productive ecosystems, perform important ecological roles. To analyze the association between Mangrove ecosystem and livelihood security of the local people, the study was conducted in Tamil Nadu, India by Sathya and Sekar (2014).They employed Garrett ranking technique to understand the positive externality of Mangrove eco-system. The shoreline protection ranked first and scored a mean value of 72. Increased fish catch and enhanced income were ranked second by the Fisherman cum farmer. Increased tourist flow and income and shoreline protection were the benefits realized by the tourist ranked first and second respectively.

Table 9.3. Positive Externalities of Pichavaram Mangrove Forest (PMF) to Fisherman-cum-Farmer and Tourist Dependents

Benefits	*Garrett's Mean Score*	*Rank*
Fisherman cum farmer		
Shoreline protection	72	I
Increased fish catch and enhanced income	60	II
Collection of NTFP	50	III
Control of soil erosion	40	IV
Water retention and purification	27	V
Tourist dependents		
Increased tourist flow and income	71	I
Shoreline protection	55	II
Control of soil erosion	45	III
Water retention and purification	31	IV

Pulse Farming and its Impact

Bhat and Umesh (2016)quantified positive externalities derived from pulse farming in the form of nitrogen fixation and its positive impact on farmers. They presented partial budgeting technique to estimate the potential benefits from pulse cultivation. To get the realistic representation, pulses growing had been compared with other crop with assumption of no added cost and reduced return instead of comparing with the land left fallow. Assuming a conservative expenditure on transportation and urea application in field and 10 per cent application losses, farmers can save 613 Rs/ha with inclusion of pulses in the farming system. Further, it would reduce the burden of government subsidy by Rs 1221/ha. After accounting for the reduction in potential damage to environment, the total positive externality of pulse farming was estimated to be Rs 2971/ha.

Besides positive externalities, negative environmental problems are very common due to effluents discharged by agro-processing industries, which has ultimately affected the livelihood of the poor. The agro-based industries cause various forms of pollution, which contaminate the surrounding air, water and land.

Pesticides and its Impact on Health

Pesticides have been an essential ally in the farmer's struggle to control different pests and constitute one of the important inputs in agricultural production. The application of pesticides has a number of benefits accompanied by disutility of negative externalities. Baba *et al.* (2017) found negative externalities due to the existence of pesticide treatment on apple in Kashmir. They found that total amount

Table 9.4. Estimating Positive Externalities of Nitrogen Fixation by Pulses-Partial Budgeting Framework

Sr. No.	*Added Cost*		**b. Reduced Cost @40kgN**	
	Items	*Value (Rs/ha)*	*Items*	*Value (Rs)/ha*
1.			**Farmers:**	
			Savings in urea application for next crop	466
			Reduction in transportation and labour costs in application of urea	100
			Savings in application losses of urea (10 per cent)	47
			Total	**613**
2.			**Government:**	
			Savings in subsidy	1221
3.			**Environment:**	
			Reduction in potential environment damages due to production and use of synthetic nitrogen fertilizers	1137
	Total (a)	0	Total (b)	2971
	c. Reduced Return		**d. Added Income**	
1.			**Improvement in soil-health:**	
			Build-up of soil-fertility through soil physical and biological properties	0
			Improvement in soil organic content	
	Total (c)	**0**	**Total (d)**	**0**
	Total (a+c)	**0**	**Total (b+d)**	**2971**
Net Income from change = Rs 2971 per ha				

spent on occupational poisoning was Rs 2211/household, of this, maximum amount was spent on neurological and gastro-intestinal problems. About Rs 86/household was spent on the treatment of cardiac problems and per case cost on this problem was more owing to costly treatment of the ailments. Due to pesticide residues, possibility of rejection of produce and contamination of water problem was reported amounted to Rs 1517/farm. Within production externalities, there were few cases of animal/poultry poisoning accounted for Rs 376/farm followed by honey bee losses *i.e.* 3550 Rs/farm and pest resistance and resurgence accounted for Rs 5500/farm respectively. Direct environmental problems resulted in loss of natural balance of ecosystem and warranted huge external cost amounted to Rs 1530/farm.

Dyeing Industry and its Impact

Devi *et al.* (2010) found negative externalities due to the pollution of Dyeing industry in Karur district of Tamil Nadu. They found the estimated values of agricultural loss and pollution averting expenditure in the Table 9.6. According to them, as the intensity of pollution increased, the value of land decreased. And as the intensity of pollution decreased, the agricultural loss also decreased. The

Table 9.5. Externalities Associated with Pesticide Treatment on Apple

Particulars	*No. of Farmers (N=200)*	*No. of Specialized Respondents (N=50)*	*External Cost (Rs/farm)*
1. Human health:			
(a) Occupational poisoning			
Respiratory	40.5	62.0	433
Neurological	58.5	68.0	880
Cardio logical	1.0	2.0	86
Dermatological	39.5	52.0	151
Ophthalmological	37.0	52.0	269
Gastro-intestinal	42.5	58.0	687
Others	0.0	1.0	5
(b) Pesticide residue			
Contamination of water, fodder, produce *etc.*	37.5	71.0	1517
2. Production externality:			
Animal/poultry poisoning	10.0	16.0	376
Honey bee loss	54.5	78.0	3550
Pest resistance and resurgence	3.5	46.0	5500
3. Environmental externality:			
Loss of beneficial insects and birds	60.5	80.0	460
Loss of biodiversity and other components of ecosystem	20.5	38.0	530
Campaign and knowledge networking, *etc.*	4.5	40.0	540

pollution averting expenditure increased with increase in the intensity of pollution. It is worth noting that the value of a highly affected farm land was less than half of a farm land less affected by pollution. To improve the quality of their land, the farmers in a highly affected area spend almost two-times of that by the farmers of less affected area in the particular district.

Table 9.6. Estimates of Land Value, Agricultural Yield Loss and Pollution Averting Expenditure in different Farm Categories in Tamil Nadu

Sl.No.	*Particulars*	*Area*		
		Less Affected	*Moderately Affected*	*Highly Affected*
1	Land value	284134	189078	135597
2	Agricultural yield loss	2859	4626	6611
3	Pollution averting expenditure	1744	2616	3534

Conclusion

Externality can be either positive or negative, prevents the attainment of Pareto-Optimality. In this review paper, positive and negative externalities and its impact on environment were documented as it differs across production systems. The positive externalities associated with livestock production in India's mixed farming system includes saving of chemical fertilizers due to use of dung as manure and prevention of carbon dioxide emission due to use of animal energy in agriculture. Mangrove ecosystem and its impact was seen in the form of positive externalities like- shoreline protection, increased fish catch and enhanced income, utilization of non-timber forest products, control of soil erosion and water retention. Besides nitrogen fixation, the positive impact of pulses was seen in the form of reduction of farmer's expenses and burden of government subsidy on fertilizers and environment because of reduced production and use of synthetic urea. The effluents discharged by the industries as well as use of pesticides, had led to severe pollution, which had ultimately affected the livelihood of the poor. It can be seen in the form of human health hazards and reducing the value of land and agricultural production on affected farms.

REFERENCES

Baba, S.H., Malik, H.A., Mir, S.A., Hamid, Y. and Kachroo, M.M. (2017). Externalities of pesticide application on apple in Kashmir Valley, *Agricultural Economics Research Review*, **30**(1): 81-92.

Bhat, S. and Umesh, K.B. (2016). Estimating positive externalities of nitrogen fixation by pulses, *Agricultural Economics Research Review*, **29**(2): 201-209.

Devi, S.T., Ramasamy, C. and Gurunathan, S. (2010). Farmers willingness to pay to avert the negative externalities of pollution of dyeing industry in Tamil Nadu, *Agricultural Economics Research Review*, **23**(2): 245-252.

Dikshit, A.K. and Birthal, P.S. (2013). Positive environmental externalities of livestock in mixed farming system of India, *Agricultural Economics Research Review*, **26**(1): 21-30.

Sathya, T. and Sekar, C. (2014). Stakeholder performance, dependence and attitude towards conservation of Mangrove eco-system in south east coast of India, *Innovare Journal of Social Sciences*, **2**(3): 15-25.

Sere, C. and Steinfeld, H. (1996). World livestock production system: current status, issues and trends, FAO Animal Production and Health Paper 127. *Food and Agriculture Organization of the United Nations*, Rome.

Swamy, M. and Bhattacharya, S. (2006). Budgeting anthropogenic greenhouse gas emission from Indian livestock using country specific emission coefficients, *Current Science*, **91**: 1340-1353.

Chapter 10

Optimization of Irrigation and Fertigation Levels for Turmeric in Western Zone of Tamil Nadu

G. Thiyagarajan[1]* and M. Manikandan[2]

[1]Water Technology Centre, Tamil Nadu Agricultural University, Coimbatore, Tamil Nadu

[2]Agricultural Engineering College and Research Institute, Tamil Nadu Agricultural University, Kumulur, Tamil Nadu

Field experiments were conducted at Agricultural Research Station, Bhavanisagar to optimize the irrigation and fertigation levels for turmeric under different lateral spacing. The experiment was laid out in split plot design with two replications and it consists of three irrigation levels of 80, 60, 40 per cent pan evaporation at a lateral spacing of 90 cm and four irrigation levels of 120, 100, 80, 60 per cent pan evaporation at a lateral spacing of 150 cm in combination with three fertigation levels of 125, 100 and 75 per cent of recommended N and K. The results revealed that drip irrigation at 40 per cent of PE with lateral spacing of 90 cm recorded lower water use (721 mm) and higher WUE (46.21 kg/ha mm). But significantly higher and comparable yield was recorded in drip irrigation at either 80 or 60 per cent of PE (37.22 and 36.5 t/ha) with lateral spacing of 90 cm and 120 per cent of PE with lateral spacing of 150 cm along with 100 per cent recommended dose of N and K (35 tha^{-1}). Economical returns were higher in 80 per cent of PE with lateral spacing of 90 cm with 100 recommended doses of N and K.

Keywords: *Drip irrigation, Fertigation, Lateral spacing, Turmeric, Water use efficiency.*

Turmeric (*Curcuma longa* L.) is one of the important spices in India with a production of 9.93 lakh tonnes from an area of 1.95 lakh ha (OGD, 2011). India is a leading country in the world in terms of turmeric producer, consumer and exporter, accounting for major share of about 78 per cent of world turmeric production and 60 per cent of the global trade (Angles *et al.*, 2011). The important turmeric growing states in India are Andhra Pradesh, Tamil Nadu, Odisha, Maharastra, Assam, Kerala, Karnataka and West Bengal.

In Tamil Nadu major turmeric sown districts are Vellore, Namakkal, Erode and Coimbatore (Thavaprakaash *et al.*, 2008). Western zone farmers of Tamil Nadu are more interested to cultivate turmeric crop due to high market value and export opportunities. Well established regulated market with canal water supply and sandy clay loam soils with good fertility also favoured the farmers to cultivate turmeric crop.

Increasing yield, dry matter production and quality of produces can be achieved if crops are grown with drip feritigation (Sharma *et al.*, 2013, Ughade and Mahadkar, 2015). Introduction of drip fertigation increases the yield of crops by 3 times and saving of fertilizer by 30 per cent (Sivanappan and Ranghaswami, 2005). Also, turmeric is a highly input responsive crop (Satyareddi and Angadi, 2014). It responds well to increased soil fertility levels (Subramanian *et al.*, 2001). Studies showed that sufficient amount of nutrients and irrigation produced higher yields with improved quality (Parthasarathy *et al.*, 2010).

Very limited research work has been carried to optimize irrigation and fertilizer levels for turmeric. Hence this study was envisaged to optimize the irrigation and fertigation levels for turmeric through drip irrigation for sandy loam soils of western zone.

Materials and Methods

Field experiment was conducted to optimize irrigation and fertigation level to turmeric for three years at Agricultural Research Station, Tamil Nadu Agricultural University, Bhavanisagar, Tamil Nadu, India. The soil of the experimental site was sandy loam with a pH of 7.5 and electrical conductivity of 0.26 dSm^{-1}. The infiltration rate of soil was 1.85 $cmhr^{-1}$, field capacity 23.1 per cent, permanent wilting point 12.5 per cent, bulk density 1.48 mgm^{-3} and the organic carbon content was 0.26 per cent. The nutritional status of the soil was low in available nitrogen (286 kg ha^{-1}) and available phosphorus (18 $kgha^{-1}$) and medium in available potassium (390 kg ha^{-1}).

The experiment consist of seven irrigation levels *viz.* drip irrigation at 80, 60 and 40 per cent of pan evaporation (PE) with lateral spacing of 90 cm (I_1, I_2, I_3) and drip irrigation at 120, 100, 80 and 60 per cent of PE with lateral spacing of 150 cm (I_4, I_5, I_6, I_7) with three fertigation levels of 125, 100 and 75 per cent of recommended N and K (F_1, F_2, F_3) was laid out in split plot design with two replications. 100 per cent recommended dose of fertilizers for turmeric is 150:60:108 NPK kg/ha. Turmeric variety used in this study was BSR2 with the seed rate of 2000 kg/ha of rhizomes. Turmeric rhizomes were sown in the bed and covering with soil.

Results and Discussion

Effect of Irrigation and Fertigation on Total Water Used and Water Use Efficiency

Different drip irrigation levels exerted significant difference on WUE of turmeric (Table 10.1). Among the different irrigation levels, drip irrigation at 40 per cent of PE (I_3) recorded lower water use (734, 797 and 632 mm) and higher WUE in all three crops (43.28, 49.31 and 46.04 kg/ha mm). It is followed by drip irrigation 60 per cent of PE (I_2). The treatments I_4 to I_7 recorded lower water use efficiencies because of increased water use. The results reported by Selvaraj *et al.* (1997) also registered higher water use efficiency in the treatment irrigating with drip at 0.36 IW/CPE ratio daily with highest yield. Thiyagarajan *et al.* (2011) reported the higher water use efficiencies in drip irrigation treatments of 80, 60 and 40 per cent PE once in 2 days as compared to surface irrigation treatment.

Table 10.1. Effect of different Irrigation Levels on Total Water Used and WUE for Turmeric

Treatment	*2013-14*		*2014-15*		*2015-16*	
	Total Water Used (mm)	*WUE (kg/ha mm)*	*Total Water Used (mm)*	*WUE (kg/ha mm)*	*Total Water Used (mm)*	*WUE (kg/ha mm)*
I_1 (90cm, 0.8PE)	1,069	31.44	1,169	38.37	863	38.46
I_2 (90cm, 0.6PE)	902	37.04	982	44.52	747	43.29
I_3 (90cm, 0.4PE)	734	43.28	797	49.31	632	46.04
I_4 (150cm, 1.2PE)	1,467	22.69	1,567	26.34	1,102	27.73
I_5 (150cm, 1.0PE)	1,299	23.85	1,406	27.67	986	29.20
I_6 (150cm, 0.8PE)	1,132	23.84	1,215	29.43	870	32.03
I_7 (150cm, 0.6PE)	974	23.74	1,033	34.61	757	34.97

Effect of Irrigation and Fertigation on Turmeric Yield

During the first-year study, among the different irrigation levels, drip irrigation at 80 per cent PE recorded significantly higher yield (33.62 t/ha) irrespective of fertigation levels. Irrespective of irrigation levels, 125 per cent recommended dose of N and K through fertigation recorded significantly higher yield (31.32 t/ha) and it is on par with 100 per cent recommended dose of N and K through fertigation. During the second, third year crop and pooled data analysis, drip irrigation at 80 and 60 per cent of PE recorded significantly higher and comparable yield. For different fertigation levels, 100 per cent recommended dose of N and K through fertigation recorded significantly higher yield. With regards to interaction, I_1F_2 (90cm, 0.8PE, 100RDF) recorded significantly higher yield for second, third crop and pooled data. It is followed by I_2F_2 (90cm, 0.6PE, 100RDF) and I_4F_1 (150cm, 1.2PE, 125RDF). A study on effect of fertigation on growth and yield of turmeric conducted by Sadarunnisa *et al.*, 2010 reported that 100 per cent recommended dose of fertilizers applied through drip recorded the maximum plant height, number of tillers per plant and fresh rhizome yield. Another study in turmeric showed that fertigation with 100 per cent levels of N and K either through straight fertilizers or through water soluble fertilizers enhances quality and to gain higher returns also (Krishnamoorthy *et al.*, 2015).

Economics of Drip Fertigation in Turmeric

The treatments I_1F_2 (90cm, 0.8PE, 100RDF) recorded the higher net return of Rs. 2,75,617, Rs. 4,20,062 and Rs. 4,09,812/ha and higher B: C ratio of 3.7, 5.6 and 5.6 during the first, second and third crop period respectively. And it was followed by I_4F_2 (150cm, 1.2PE, 100RDF) and I_2F_2 (90cm, 0.6PE, 100RDF). The treatments in I_1(90cm,0.8PE), I_2 (90cm,0.6PE), I_3 (90cm,0.4PE), I_4 (150cm,1.2PE), I_5 (150cm,1.0PE) recorded B: C ratios higher than 4.

Yield increase of 90 cm lateral spacing over 150 cm lateral spacing was only 6 per cent. Turmeric raised at 150 cm lateral spacing has performed equally well with turmeric raised at 90 cm lateral spacing. Moreover, yield increase of 90 cm lateral spacing over 150 cm lateral spacing was only 6 per cent. But from the economics point of view 150 cm lateral spacing with 120 per cent PE may be feasible.

Conclusion

Based on the results of the study, it can be concluded that drip irrigation at 40 per cent of PE with lateral spacing of 90 cm for lower water use and higher WUE. Drip irrigation at 80 per cent of PE with lateral spacing of 90 cm and 120 per cent of PE with lateral spacing of 150 cm with 100 per cent recommended dose of N and K through fertigation for higher yield and economics.

REFERENCES

Angles, S., Sundar, A. and Chinnadurai, M. (2011). Impact of globalization on production and export of turmeric in India – An economic analysis. *Agric. Eco. Res. Rev.*, 24: 301-308.

Krishnamoorthy, C., Soorianathasundaram, K. and Mekala, S. (2015). Effect of fertigation on FUE, quality and economics of cultivation in turmeric (*Curcuma longa* L.) cv. BSR 2. *International Journal of Agricultural Science and Research*, 5(1): 67-72.

OGD (Open Government Data) Platform India (2011). Area/Production and Productivity of various kinds of Spices (Assessed on 07-09-2016)

Parthasarathy, V.A., Dinesh, R., Srinivasan, V. and Hamza, S. (2010). Integrated nutrient management in major spices. *Indian J. Fert.*, 6: 110-128.

Sadarunnisa, S., Madhumathi, C., Srinivasa Rao, G. and Sreenivasulu, B. (2010). Effect of fertigation on growth and yield of turmeric cv. Mydukur. *J. Hortl. Sci.*, 5(1): 78-80.

Satyareddi, S.A. and Angadi, S.S. (2014). Response of turmeric (*Curcuma longa* L.) varieties to irrigation methods and graded levels of fertilizer. *Res. Environ. Life Sci.*, 7(4): 237-242.

Selvaraj, P.K., Krishnamurthi, V.V., Manickasundaram, P., Martin, G.J. and Ayyaswamy, M. (1997). Effect of irrigation schedules and nitrogen levels on the yield of turmeric through drip irrigation. *Madras Agric. J.*, 84: 347-348.

Sharma, S., Sanmay Patra, Gokul B. Roy and Soumen Bera (2013). Influence of drip irrigation and nitrogen fertigation on yield and water productivity of guava. *The Bioscan.*, 8(3): 783-786.

Sivanappan, R.K. and Ranghaswami, M.V. (2005). Technology to take 100 tons per acre in Sugarcane. *Kissan World*, 32: 35-38.

Subramanian, K.S., Sivasamy, N., and Thangaraj, T. (2001). Integrated nutrient management for turmeric. *Spice India*, 14(12): 25-26.

Thavaprakaash, N., Babu, C. and Jagannathan, R. (2008). Identifying potential cropping zones for important horticultural crops of Tamil Nadu. *Madras Agric. J.*, 95(7-12): 418-424.

Thiyagarajan, G., Vijayakumar, M., Selvaraj, P.K., Duraisamy, V.K. and Mohamed Yassin M. (2011). Performance evaluation of fertigation of N and K on yield and water use efficiency of turmeric through drip irrigation. *International Journal of Biostress Management*, 2(1): 69-71.

Ughade, S.R. and Mahadkar, U.V. (2015). Effect of different planting density, irrigation and fertigation levels on growth and yield of Brinjal (*Solanum melongena* L.). *The Bioscan*, 10(3): 1205-1211.

Chapter 11

Effect of Organic Sources of Nutrients in Yield, Soil Health and Economics of Vegetable Crop for Maintaining Sustainable Agriculture in RiBhoi District of Meghalaya, North-east India

Popiha Bordoloi

Subject Matter Specialist, Krishi Vigyan Kendra Ri-Bhoi, ICAR (RC) for NEH Region, Umiam – 793 103, Meghalaya

Agriculture is the backbone of poverty reduction as well as socio-economic development of Northeast India. But the application of organic as well as inorganic fertilizer has been very limited among the farmers of RiBhoi District of Meghalaya. The farmers get lower yield due to improper fertility management. Due to mono-cropping as well as imbalance fertilizer application to the crop field, soils are also deteriorated day by day in some cases. A Front Line Demonstration was conducted at the farmers' field to demonstrate the integrated nutrient management for tomato crop for maintaining the soil health, reducing the rate of chemical fertilizer and to increase the yield of tomato. The demonstration was conducted at ten different farmers' field of RiBhoi District of Meghalaya during the year 2015-16 and 2016-17. The area under each demonstration was 0.4 ha. The treatment comprise of Azotobacter and PSB 2 kg each by root dip treatment + vermicompost @ 2.0 t/ha along with 50 per cent recommended dose of chemical fertilizer and Azotobacter +PSB @ 2kg each + 30 kg cow dung in 5 liter water/ha by root dip treatment for 30 min and the farmers' practice. The results of the FLD revealed that the application of Azotobacter and PSB 2 kg each by root dip treatment + vermicompost @ 2.0 t/ha along with 50 per cent recommended dose of chemical fertilizer had given significantly higher yield i.e. 241 q/ha and B.C ratio of 1.71 followed by Azotobacter +PSB @ 2kg each + 30 kg cow dung in 5 liter water/ha by root dip treatment for 30 min (204 q/ha yield, B.C ratio 1.52) and the farmers practice (168 q/ha yield, B.C ratio 1.29). Moreover, soil health was also improved with the increased amount of soil nutrients at the time of harvesting as compared to farmers practice.

Keywords: *FLD, Tomato, Soil health, Organic farming, Sustainable agriculture.*

Agriculture is the most important enterprise among all other human enterprises for poverty reduction in Northeast India. For maintaining sustainable crop production as well as to reduce the cost of fertilizers application, the organic residue that arises out of the agricultural system itself need to be recycled. After harvesting the crops, various residues left in the field are either burnt or undergoes natural degradation process that helps in soil organic matter amendment. It is also required to maintain the relationships between organisms and their environment and principles of ecology should maintain for sustainable agriculture. The productivity and efficiency of agro-ecosystems depends on the quality of soil organic matter and nutrient mineralization pattern, which is affected by various management practices during the cropping and post-cropping phase in the field (Ramakrishnan, 1994). Application of vermicompost which could be achieved by recycling of organic waste in the crop field increases the crop yield to a desired level. It also increases the physical, chemical and biological properties of soil and hence improves soil productivity. Many research reports do show that crop residues applied to a crop has benefited to the succeeding crop. Anon (1996) reported that continuous application of farm yard manures at 10 t ha^{-1} was responsible to increase the yield of finger millet by 1447 kg ha^{-1} over absolute control and 152 kg ha^{-1} over continuous application of NPK alone. Wagger *et al.* (1987) and Ladd *et al.* (1983) have also shown that in Wheat-Sorghum and Alfalfa-Wheat systems, 12-33 per cent and 20-30 per cent N mineralized by the crop residue applied to first crop becomes available for the succeeding crop.

The use of chemical fertilizers alone has deteriorated the soil physical, chemical, biological and microbial properties. It provides only specific nutrients needed by the plant. Moreover, it is a costly input and sometimes farmers have to face difficulties due to its lake of availability in proper time and proper amount. On the other hand, the application of organic manures alone has the drawback of low content of plant nutrient and its slow nutrient release characteristics. So integrated nutrient management including both organic and inorganic fertilizers will be more profitable where the supply of both organic and inorganic fertilizers is not constraints. The improper application of nutrient requirement by the crop by organically or by inorganically is the major cause of low productivity of vegetable crop of Meghalaya. Tomato is an important vegetable crop of Meghalaya and maximum of the farmers of RiBhoi District are engaged in Tomato cultivation in Rabi season. Tomato grown in rain fed area is faced with water scarcity, temperature extremes, irregular rainfall pattern and minimal use of nutrients that limit the yield potential and also results in irregular production. The productivity of Tomato per unit area could be increased by adopting recommended scientific and sustainable management production practices using suitable high yielding varieties. Front line demonstrations (FLD) were carried out in a systematic manner at farmers' Field to convince farmers to adopt improved production management practices for enhancing Tomato productivity.

Materials and Methods

A survey was conducted on three villages of RiBhoi District namely Liarkhla, Kdonghulu and Umroi to explore the detail information of study area. Secondary data were collected from Department of Agriculture, RiBhoi, ICAR (RC) for NEH Region, Umiam Krishi Vigyan Kendra RiBhoi, District Library, Different Journal, books and from Internet. The demonstration was conducted at ten different farmers' field of RiBhoi District of Meghalaya during the year 2015-16 and 2016-17. The area under each demonstration was 0.4 ha. The treatment comprise of Azotobacter and PSB 2 kg each by root dip treatment + vermicompost @ 2.0 t/ha along with 50 per cent recommended dose (60:40:30) of chemical fertilizer. Farmers' practice include imbalance fertilizer application (30:20:20 NPK kg/ha) without seed treatment. All the participating farmers were trained on all aspects of tomato production and soil fertility management before implementing the FLDs at their field. To study the impact of frontline demonstration, data from FLDs and local practices were collected and analyzed. The extension gap, technology gap and technology index along with benefit cost ratio were calculated using the formula as suggested by Samui *et al.* (2000) as follows

Technology Gap = Potential yield – Demonstration Yield (q/ha)

Extension Gap = Demonstration Yield – Farmers Yield (q/ha)

Technology Index = Potential Yield – Demonstration Yield/Potential Yield x 100

The soil fertility statuses were estimated by soil analysis of composite soil sample from each plot before transplanting and after harvesting of crop. The soil of the experimental site was sandy loam and acidic in reaction. Data were analyzed using the standard statistical packages and some descriptive statistics to interpret the data.

Results and Discussion

Seed Yield

The application of Azotobacter and PSB 2 kg each by root dip treatment + vermicompost @ 2.0 t/ha along with 50 per cent recommended dose of chemical fertilizer had given significantly higher yield *i.e.* 241 q/ha followed by *Azotobacter* +PSB @ 2kg each + 30 kg cow dung in 5 liter water/ha by Root dip treatment for 30 min (204 q/ha) and the farmers practice (168 q/ha) (Table 11.1). Similar results have een reported by Joshi *et al.* (2015).

Economics

The Net Return and Benefit Cost ratio were calculated by taking the prevailed cost of commodities during both the year the year of demonstration. From the Table 11.1 it is seen that the Net Return is highest in the application of *Azotobacter* and PSB 2 kg each by root dip treatment + vermicompost @ 2.0 t/ha along with

Azotobacter and PSB 2 kg each by root dip treatment + vermicompost @ 2.0 t/ha along with 50% recommended dose (60:40:30) of chemical fertilizer

Transplanting of Tomato seedling after Root Dip treatment with bio fertilizer

Azotobacter and PSB 2 kg each by root dip treatment + vermicompost @ 2.0 t/ha along with 50% recommended dose (60:40:30) of chemical fertilizer

50 per cent recommended dose of chemical fertilizer *i.e.* Rs. 150097/-. This is also recorded that the application of Azotobacter and PSB 2 kg each by root dip treatment + vermicompost @ 2.0 t/ha along with 50 per cent recommended dose of chemical had given significantly higher B.C ratio of 1.71 followed by Azotobacter +PSB @ 2kg each + 30 kg cow dung in 5 liter water/ha by Root dip treatment for 30 min (B.C ratio 1.52) and the farmers practice (B.C ratio 1.29).

Extension Gap and Technology Gap

The extension gap in the Azotobacter and PSB 2 kg each by root dip treatment + vermicompost @ 2.0 t/ha along with 50 per cent recommended dose of chemical fertilizer was found 73 q/ha and for the treatment Azotobacter +PSB @ 2kg each + 30 kg cow dung is found 36 q/ha emphasized the need to trained the farmers through demonstration and training for adoption of improved technology to reverse the trend of wide extension gap (Table 11.1). Technology gap of 9 q/ha reveals the farmer's cooperation in carrying out new demonstration with good results. The technology index of 3.6 per cent showed the feasibility of the demonstrated technology at the farmers' field and suitability of the Technology in the region.

Soil Fertility Status

From the Table 11.2 it is found that soil fertility status was increased at the demonstrated plot. The soil samples were taken at the time harvesting for all the plots under demonstration as well as from the Farmers practice plot which was under supervision. The data revealed that application of Azotobacter and PSB 2 kg each by root dip treatment + vermicompost @ 2.0 t/ha along with 50 per cent recommended dose of RDF recorded highest amount of organic carbon *i.e.* 1.39 kg/ha compare to control plot 1.12 kg/ha. Similar results also found for available phosphorus and available potassium (Table 11.2) in both the treatment for demonstration compare to control plot. So, soil nutrient status can be improved by adapting the demonstrating technology.

Conclusion

Tomato is very popular crop among the farmers of RiBhoi District of Meghalaya. From the study, it can be concluded that application of Azotobacter and PSB 2 kg each by root dip treatment + vermicompost @ 2.0 t/ha along with 50 per cent

Table 11.1. Effect of using INM on Yield and Economics of Tomato under Rain Fed Condition

Technology Details	*Yield*			*Per cent Increase/ Change in Avg. Yield Over Local*	*Gross Cost (Rs/ha)/ (Rs./ unit)*	*Gross Return (Rs/ha)/ Rs./unit)*	*Net Return (Rs/ha)/ (Rs./ Unit)*	*B:C Ratio (GR/ GC)*	*Technology Gap (q/ha)*	*Extension Gap (q/ha)*	*Technology Index*
	High	*Low*	*Avg. yield (q/ ha)*								
T1: Azotobacter and PSB @ 2 kg each by root dip treatment + vermicompost @ 2.0 t/ha + 50 per cent RDF (60:40:30)	286	196	241	143.45	211403	361500	150097	1.71	9	73	3.6
T2: Azotobacter +PSB @ 2kg each + 30 kg cow dung in 5 liter water/ha by root dip treatment for 30 min	163	212	204	121.43	201504	306000	104496	1.52	46	36	18.4
T3: Farmers Practice (30:20:20 NPK kg/ha)	180	126	168	-	195348	252000	56652	1.29	-	-	-

Table 11.2. Effect of using INM on Soil Fertility Status under Rain Fed Condition

Technology Details	*pH*	*Organic Carbon (kg/ha)*	*Available Nitrogen (kg/ha)*	*Available Phosphorus (kg/ha)*	*Available potassium (kg/ha)*
T1: Azotobacter and PSB @ 2 kg each by root dip treatment + vermicompost @ 2.0 t/ha + 50 per cent RDF (60:40:30)	4.93	1.39	386.48	52.28	169.61
T2: Azotobacter +PSB @ 2kg each + 30 kg cow dung in 5 liter water/ha by root dip treatment for 30 min	4.85	1.23	326.20	47.54	143.26
T3: Farmers Practice (30:20:20 NPK kg/ha)	4.67	1.12	305.35	41.24	121.54

recommended dose of RDF was found effective for getting the highest yield of Tomato as well as for maintaining the sustainability in Agriculture. The suitable BC ratio reveals the economic viability of the demonstration and convinced the farmers to adopt the Technology. The soil fertility status also improved by the use of the Treatments from the above Technology. So, this soil test based Technology can defiantly play an important role for motivating the farmers for adoption of new technology by popularizing the FLD.

REFERENCES

Anonymas (1996). Annual Report of dry land Agriculture Project, USA, Bangalore. 28-57.

Joshi N S, Bariya M K and Kunjadia B B (2015). Yield gap analysis through front line Demonstration in Wheat crop.

Ladd, J. M., Amalo, M., Kackson, R. B. and Butter, J. H. A. (1983). Utilization by wheat crop of N from legume residue decomposing in the soil in the field. *Soil Biol. and Bio. Chem.* **15**: 231-238.

Ramakrishnan, P. S. (1994). The *Jhum* agro-ecosystem in north-eastern India: A case study of biological management of soils in a shifting agricultural system. *The Biological Management of Tropical Soil Fertility*; Eds P. L. Woomer and M. J. Swift, Wiley-Sayce Publications, Chichester. 189-207.

Wagger, M. G., Kissel, D. E. and Smith, S. S. (1987). Mineralization of Nitrogen from N^{15} labeled crop residue under field condition. *Soil Sci. Soc. Am. J.* **49**: 1220-26.

Chapter 12

Agricultural Diversification: An Environment Friendly Approach for Sustainable Production

Sumit Sow*[1]*, Shivani Ranjan and Sarita

PG Scholar, Department of Agronomy, Bihar Agricultural University, Sabour, Bhagalpur – 813 210, Bihar

Agriculture has always been the backbone of the Indian economy, but with globalization of market, shrinking land and climate change, increasing population; crops should be produced sustainably. So, agricultural diversification is an essential strategy, which enhance agricultural income, minimizes risks due to crop failures and helps to earn foreign exchange. It can be designed to reduce poverty, employment generation and environmental conservation. There has been a marked decreased of area in cereals and an increase in pulses, oilseeds and cash crops which are comparatively higher value crops. Moreover, the development of livestock sector is more inclusive and can result in a sustainable agriculture system. There have been noteworthy changes in the pattern of agricultural diversification at the regional level. Moreover, agricultural diversification increases both individual and social gain. Hence, it has become a most popular concept attracting the agrarians, scholars and economists. Thus, agricultural diversification is as an alternative for maximizing crop productivity to ensure food security and enhancing the environment sustainability for better livelihood.

Keywords: *Agricultural diversification, Climate change, Crop productivity, Environmental conservation, Food security.*

The Indian Economy has undergone structural changes overtime with the anticipated decline in the share of agriculture in GDP, agriculture has not diminished for two major reasons; first, the country achieved self-sufficiency in food production at the macro level but still is a food deficit country facing massive challenges of food security and high incidence of rural poverty, pressure on agriculture for production and raise farmer's income is high; second, the reliance of the rural workforce on agriculture for employment has not declined with respect to the sectoral commitment to GDP. As the cultivated area remains more or less constant, but increase in food demand because of rise in population and urbanization puts agricultural land under pressure resulting in crop intensification and interchange of food crops with commercial crops.

Agricultural diversification has emerged as an important alternative to attain the objectives of output growth, employment generation and natural resources, sustainability in the developing countries. It is a movement of production-portfolio from a low-value commodity mix (crop and livestock) to high-value commodity mix (crops and livestock). The focus is on the horticulture, dairy, poultry and fisheries sectors. These are perishable in nature but yield high, quick and regular dividends to the farmers. These commodities are also labour-intensive, which is the strength of small farm holders, as they possess large reserves of unemployed or underemployed family labour.

GDP Trend of Agriculture

Agricultural diversification in India is gradually picking momentum in favour of not only high value crops but also livestock activities to augment incomes rather than a coping strategy to manage risk and uncertainty. Crops, livestocks, fisheries and forestry constitute the core sub-sectors of agriculture. Crop sub-sector is the principal source of generating income in agriculture followed by the livestock sub-sector (Table 12.1). The table shows trend in area under major crops *i.e.*, Rice, Wheat, Nutri Cereals, Pulses, and Food grains, Oilseeds, Sugarcane and Cotton.

Agricultural Diversification and its Components

Generally, the former type of diversification indicates shift from one crop to another crop or from one enterprise/sub-sector to another enterprise/sub-sector, the other type of diversification may involve income-enhancing enterprises in addition to the existing ones. In the agricultural scenario, diversification can be regarded as the re-allocation of farm's productive resources such as land, capital, farm equipment and pieces to other farmers and, particularly in richer countries, non-farming activities which can increase incomes in agriculture, minimize risks due to crop failures and above all, earn foreign exchange. In India, diversification has occurred across the crop as well as within the crop, livestock, and forestry and fishery sectors.

Two approaches to crop diversification in agriculture:

a) Horizontal diversification- The primary approach to crop diversification used in production agriculture. In this approach, diversification normally occurs through crop intensification which means addition of new high-value crops to existing cropping systems as viable way of improving the overall productivity of a particular farm or farming economy of a region as a whole.

Table 12.1. Area under Major Crops 1950-51 Onwards (in million hectares)

Crops	1950-51	1960-61	1970-71	1980-81	1990-91	2000-01	2010-11	2011-12	2012-13	2013-14	2014-15	2015-16	2016-17*
(1)	(2)	(3)	(4)	(5)	(6)	(7)	(8)	(9)	(10)	(11)	(12)	(13)	(14)
Rice	30.81	34.13	37.59	40.15	42.69	44.71	42.86	44.01	42.75	44.14	44.11	43.50	43.19
Wheat	9.75	12.93	18.24	22.28	24.17	25.73	29.07	29.86	30.00	30.47	31.47	30.42	30.60
Nutri Cereals	37.67	44.96	45.95	41.78	36.32	30.26	28.34	26.42	24.76	25.22	25.17	24.39	24.77
Pulses	19.09	23.56	22.54	22.46	24.66	20.35	26.40	24.46	23.26	25.21	23.55	24.91	29.46
Foodgrains	97.32	115.58	124.32	126.67	127.84	121.05	126.67	124.75	120.78	125.04	124.30	123.22	128.03
Oilseeds	10.73	13.77	16.64	17.60	24.15	22.77	27.22	26.31	26.48	28.05	25.59	26.08	26.20
Sugarcane	1.71	2.42	2.62	2.67	3.69	4.32	4.88	5.04	5.00	4.99	5.06	4.92	4.38
Cotton	5.88	7.61	7.61	7.82	7.44	8.53	11.24	12.18	11.98	11.96	12.81	12.29	10.84
Jute & Mesta	0.57	0.90	1.08	1.30	1.02	1.02	0.87	0.90	0.86	0.83	0.81	0.78	0.76

* 4th Advance Estimates

Source: Department of Agriculture, Cooperation & Farmers Welfare

Growth (over the previous year) in the GDP of agriculture is given below:

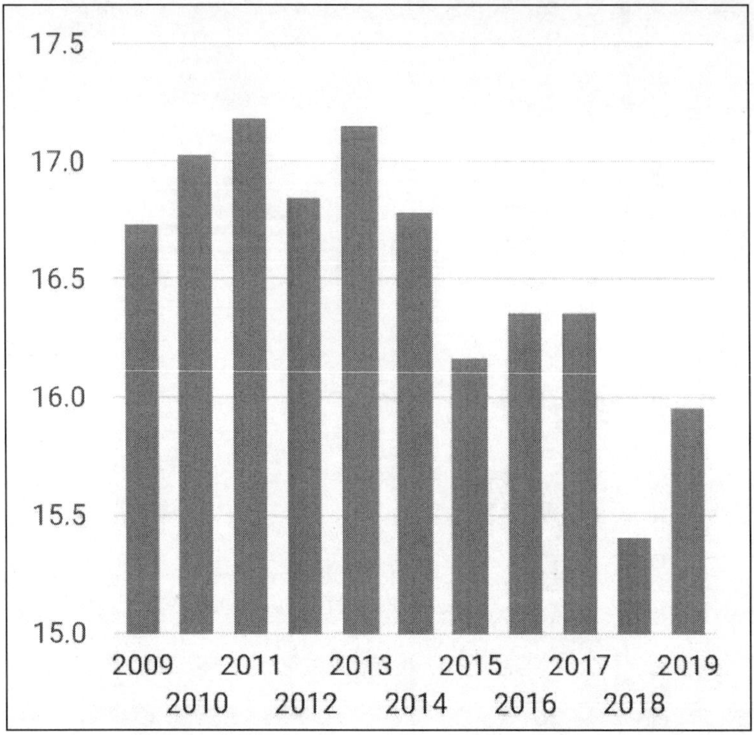

*Source: Central Statistics Office .

b Vertical diversification- It is an approach in which value is added to the products by farmers through various methods such as processing, regional branding, packaging, merchandising, or other efforts to enhance the product.

Need of Agriculture Diversification

A considerable portion of Indian farming is dependent on the vagaries of rainfall, making it is dangerous issue to rely upon exclusively. Accordingly, the need of diversification is required to enable the farmers to earn from other alternatives non-farm occupation (Gupta *et al.*, 2013). Due to overcrowding agriculture cannot further generate employment opportunities. Therefore, the prospects of the non-farm sector ought to be opened up in the rural area to provide job openings for work, diverting workforce from the already crowded agriculture sector. The non-farm sector has several segments that possess dynamic linkages such enhance the healthy growth of an economy.

The important facts highlighting the need of crop diversification in India are-

- ☆ For raising farm income.
- ☆ Sustainable production and income and nutrient security.

- ☆ Promotion of export.
- ☆ Employment generation and poverty alleviation.
- ☆ Judicious use of land and water resources.
- ☆ Reduction in environmental pollution.

Major Driving Forces for Crop Diversification

- ☆ Increasing income on small farm holdings.
- ☆ Withstanding price fluctuation, responding to changing consumer demands
- ☆ Mitigating ill-effects of aberrant weather and climate change.
- ☆ Food demand, improving fodder for livestock animals.
- ☆ Conservation of natural resources (soil, water, *etc.*).
- ☆ Minimizing environmental pollution and dependence on off-farm inputs.
- ☆ Insect pests, diseases and weeds problem.

Horticulture Based Agricultural Diversification

Fruits, vegetables and similar exportable crops often termed as 'high value' crops in the ongoing diversification play an important role in growth of Indian agriculture. The demand-side pull for fruits and vegetables was further strengthened with the opening up of the economy and increase in per capita income (Joshi *et al.*, 2007). The prices of fruits and vegetables have increased consistently. During the past two decades, area and production of vegetables increased considerably in India. Besides, it is emerging as an important source of income augmentation for small farm holders in water-scarce regions due to massive subsidies extended by the government on water-saving devices (*e.g.*, sprinkler and drip systems). In addition, the watershed programmes gave high priority to vegetable production to enhance the efficiency of scarce water conserved in the rainfed areas. Indian mangoes specially Dasheri, and Alphonso have excellent export market. Dry fruits and spices have also gained during the past two decades (Sharma, 2005).

Crop Based Agricultural Diversification

Crop diversification is not just move from conventional and less remunerative to more remunerative crops, but it is demand driven, situation specific and dynamic concept involves spatial, temporal and value addition, resource complementary approach. It implies the use of environmental and human resources to grow a mix of crops with complimentary marketing opportunities and a shifting of resources from low value crops to high value crops. Non-foodgrain crops, like oilseeds, fibre crops, potato, sugarcane have mainly substituted for coarse cereals in search of higher incomes. Cereals dominated foodgrain crops and area and production of rice, wheat and,maize are rising and those of barley, millets and sorghum were rapidly decreasing. Maize is emerging as an important crop mainly to meet the requirements of booming poultry sector and availability of improved hybrids, flexibility in growing seasons and diverse uses of maize are responsible for its area expansion. It has been recommended that India should diversify its agriculture and get a traction on the world market (Radhakrishna and Reddy, 2004). Consequences of changes in cropping pattern are:

1. The developments have positive effects on land/labor efficiency and net farm income, yet it has also imperiled a various undesirable side effects like diminished farm employment and crop imbalances and loss of biodiversity at the farm level.
2. Although the expansion of commercialized agriculture has fomented not only new sets of rural non-farm activities but also strengthened the rural-urban growth linkages, but it has also weakened the conventional inter-sectoral linkages between the crop and livestock sectors.
3. Crop pattern changes lead to serious environmental consequences that take such forms as groundwater depletion and water logging and salinity which can reduce the productive capacity over the long term.
4. Leguminous crop is taken in crop rotation with non-leguminous crop to maintain soil fertility as well as less use chemical fertilizer.

Livestock Led Diversification

Livestock output in India, is growing faster than any other agricultural sub-sector and accounted for less than one-fifth of agricultural output in the early seventies; the corresponding figure has increased to 40 per cent in the recent years (after 2000s). This is often considered as a new source of agricultural growth in the country. There has been supply as well as demand side impetus for growth of dairy in the livestock sector in India. Livestock products have become increasingly significant in the food basket of consumers. Livestock-based rural livelihoods have emerged as important in India with the increased fragmentation of land and increased number of small and marginal farmers. In India, mixed farming has been an viable option and in such a system, agriculture and livestock have a integral relationship. This complementary relationship that thrives with the use of inputs from one sub-system to another is weakening with the onslaught of commercialization. Livestock sector has been growing at a fast rate and its share in total value of agricultural output has been dynamically boosting in India (Birthal and Parthsarthy, 2002). Future of livestock sector is quite promising in the country, as there is still huge potential to augment production, consumption and export of different livestock commodities.

Diversification within the Fisheries Sector

Fisheries sector has also diversified over the years. It was mainly due to gradual shift from marine to inland fisheries. The inland fish potential is still higher with a well spread location of rivers, canals and reservoirs. The higher growth in inland fisheries is mainly attributed to the overwhelming progress in aquaculture, both in fresh and brackish waters. The share of culture fisheries in the inland sector has ascended from about 43 per cent in 1984-85 to a significant level of about 84 per cent in 1994-95 (Kumar *et al.*, 2001). A bulk of growth in culture fisheries has come from the fresh water aquaculture (Krishnan *et al.*, 2000). The expansion of inland fisheries has also led to some negative externalities related to degradation of arable lands due to salinity. Several production and development oriented programmes were launched such as Development of Freshwater Aquaculture, Integrated Coastal Aquaculture, and Development of Coastal Marine Fisheries. To encourage the aquaculture, the programmes were started to upgrade the technology, and to encourage involvement of private sector for different activities like quality seed, feed and other inputs and formation of suitable infrastructure for storage, transport, marketing and credit. To build up better infrastructure facilities, 'Fisheries Industrial Estates' were created by grouping the cluster of fishing villages. There exists a promising export market for marine and inland fish along with aqua products.

Integration of Consumption and Production

A majority of Indian farmers are smallholders. Their transaction costs are higher due to (i) lack of access to markets, (ii) limited marketable surplus, and (iii) perishable nature of the product. Due to rising demand for high-value commodities, different forms of production-market integration are slowly emerging in the food supply chain. These include (i) spot or open market transactions, (ii) agricultural co-operatives, and (iii) contract farming. Each mode of integration has some advantages, and the success depends upon the interest of all the parties.

Challenges in Crop Diversification

The major constraints are:

1. Lack of suitable land because most of the cropped area in the country is completely dependent on rainfall.
2. Sub-optimal and over-use of resources like fertilizers, land and water, causing negative impact on the sustainability of agriculture.
3. Fragmentation of land, less favoring modernization and mechanization of agriculture.
4. Very weak agro-based industry and inadequate research - extension - farmer linkages inadequately trained human resources and lack of knowledge among the farmers.
5. Inadequately trained human resources together with persistent and large scale illiteracy among farmers.
6. Decreased investment in agriculture sector over the years.
7. Low adoption rate of new varieties and inadequate supply of improved and quality seeds and planting material of improved cultivars.
8. Existing marketing system as well as import discourages diversification.
9. Inadequate post-harvest technologies and inadequate infrastructure for post-harvest handling of perishable horticultural produce.

Government Policies for Agricultural Diversification

- ✰ Implementing national agriculture insurance scheme.

- Technology mission on cotton.
- Provision of capital subsidy of 25 per cent for construction/modernization/ expansion of old storages and storages for horticultural produce.
- Creation of watershed development fund.
- Strengthening agricultural marketing.
- Seed crop insurance and seed bank scheme.
- Cooperative sector reforms.

Strategies and Policy Options for Agricultural Diversification

The achievement of any diversification programme includes the development of crop or enterprise specific technologies, recognition and development of market, and provision of different economic incentives. The free trade offers the movement from closed to market economies and leaves it to the market, domestic or international, to judge the competitive advantage of a specific product (Birthal *et al.*, 2007). Appropriate policy support for attracting private sector in strengthening farm-firm integration would enhance supply of high-value commodities, and benefit the small farm-holders. In terms of policy implications and strategy, the following points are worth considerations:

- In policymaking, greater thrust should be given to non-grain economy in terms of R and D expenditure, and investments in marketing, storage and processing facilities.
- The government needs to enhance the allocation of resources to the basic research on these high-value commodities, as well as their marketing, storage and processing. The private sector, including foreign direct investment (FDIs), has to be invited in a big way.
- To lay an efficient foundation for value-addition processes, it is necessary to withdraw preferences extended to cooperatives and public sector concerns in the form of corporate tax exemptions, subsidised finance, *etc.*
- It is essential to facilitate the emergence of vertical integration between farmers, processors and retailers (farm-firm-fork linkages) in high-value agriculture. To ensure food safety, laws need to be duly enforced, and sanitary and phyto-Sanitary (SPS) standards be adopted and the fruits of the innovative technologies should reach the farmers at the earliest.
- To sustain and operationalize crop diversification, natural Research Management based Institutions and State Agricultural Universities should jointly addressing the issues connected with the agricultural diversification.
- Establishment of Crop Directorates for each of the major crops and groups of crops for development and technology transfer focus on each of these crops and commodities.

Conclusions

The major concern of Indian agriculture has been on fostering production of cereals, particularly wheat and rice. It has been observed that production of horticultural commodities, milk, meat, fish and eggs, these areas under high-value crops have recorded significant growth (1.99 per cent) during the past two decades. Consequently, the share of high-value crops in the total value of agricultural output has noticeably enhanced. Strengthening of horticulture, livestock and fisheries sectors can significantly contribute to the enhancing of farm income, offering employment opportunities in rural areas and meeting the food and nutritional needs of rural poor. Indian agriculture has witnessed diversification with quite amazing improvements in the share of both livestock and fisheries sector in the total income from agriculture. Within the crop sector, the diversification has largely been in favour of non-foodgrain crops in most of the states. However, the diversification in these states has not been essentially for income generation, but has also been for mitigating risk. To cater the increasing demand for food, agricultural diversification is a viable option which not only increase the crop productivity but also maintain the environmental sustainability.

REFERENCES

Birthal, Pratap S. and P. Parthasarathy Rao (eds). (2002). Technology Options for Sustainable Livestock Production in India. National Centre for Agricultural Economics, and Policy Research, New Delhi and International Crops Research Institute for the Semi-Arid Trop'cs, Parancheru, India.

Birthal, P.S., P.K. Joshi, Devesh Roy and Amit Thorat (2007), Diversification in Indian Agriculture Towards High Value Crops: The Role of Small Holders, IFPRI Discussion Paper # 00727, November, International Food Policy Research Institute, Washington D.C., U.S.A.

Government of India (2017). Pocket Book of Agricultural statistics, Ministry of Agriculture and Farmers Welfare Department of Agriculture, Cooperation and Farmers Welfare Directorate of Economics and Statistics, New Delhi.

Gupta H (2013) Agricultural diversification in India. *International Journal of Advanced Research in Management and Social Sciences*, 2(5): 221–231.

Joshi, P.K., Ashok Gulati and P.S. Birthal (2007), "Agricultural Diversification in India: Status, Nature and Pattern in India", in P.K. Joshi, Ashok Gulati and Cummings Jr. (2007) (Eds.), Agricultural Diversification and Smallholders in South Asia, Academic Foundation, New Delhi.

Krishnan, M. and PS Birthal (eds) (2000). Aquaculture in IndIa: Retrospect and Prospect, National Centre for Agricultural Economics and Policy Research (NCAP), New Delhi.

Kumar, Anjam, P.K. Joshi and P. S. Birthal (2001). "Fisheries Sector in India: An Overview of Performance, Policies and Programs", 10 International Workshop on Strategies and Options for Increasing and Sustaining FIsheries Production to Benefit Poor Households in Asia, August 20-25, ICLARM. Penang, Malaysia.

Kumar, Ranjit and Singh, Naveen and Singh, R.P. (2006). Diversification of Indian Agriculture: Composition, Determinants and Trade Implications. *Agricultural Economics Research Review*. 19.

Radhakrishna, R. and K.V. Reddy, (2004) Food security and nutrition: Vision 2020,India Vision 2020, Report of the Committee on India Vision 2020, Planning Commission, Government of India, Academic Foundation.

Sharma, H.R. (2005), "Agricultural Development and Crop Diversification in Himachal Pradesh: Understanding the Patterns, Processes, Determinants and Lessons", *Indian Journal of Agricultural Economics*, Vol. 60, No. 1, January-March, pp.71-93.

Chapter 13

Integrated Plant Nutrient Management for Sustainable Agriculture

A.K. Mishra[1], *Ravindar Kumar*[2], *S.K. Mishra*[3] *and Sheesh Pal Singh*[1]

[1]*Krishi Vigyan Kendra, Amroha , Uttar Pradesh*
[2]*Krishi Vigyan Kendra, Muradabad-II, SVPUAT, Meerut, Uttar Pradesh*
[3]*Krishi Vigyan Kendra, DUVASU, Mathura , Uttar Pradesh*

Integrated plant nutrient supply involves monitoring all the path ways of plant nutrient supply in crops and cropping systems and calls for judicious combination of fertilizer, bio-fertilizer and organic manures. Organic sources of plan nutrients including growing of legumes in cropping systems, green manures, crop residues, organic manures (FYM, compost, Vermicompost, biogas slurry, phosphor compost, bio compost, press mud, cakes etc.) and bio-fertilizers. The available information has shown that by addition of organic manures in addition of fertilizers (add-on series) good crop yields were sustained over long periods as compared to a decline in crop yields when only fertilizers were applied, especially when N alone was applied. The data from replacement series of trials shown that in most cropping systems specially in rice wheat, rice wheat sugarcane systems application on 50 per cent N through green manures, FYM or crop residues or 50 per cent recommended dose of fertilizer (RDF) (NPK) to Kharif rice and 100 per cent recommended dose of fertilizer to Rabi crop (wheat) gave the same yield as obtained with 100 per cent recommended dose of fertilizer of Kharif and Rabi crops (rice/wheat). These results show that 25 per cent NPK applied to the cropping system can be saved. Further addition of organic manures both in add-on and replacement series always improved chemical, physical and biological properties of soils. As regards bio-fertilizers usefulness of rhizobial inoculation of pulses and leguminous oilseeds (soya bean, groundnut) along with recommended dose of fertilizer (NPK) has been proved beyond doubt. Rhizobial inoculation helps not only in increasing crop yields but also increasing the amount of an left after succeeding crops use of cultures of PSO's [phosphate solubilising bacteria (PSB) and phosphate solubilising fungi (PSF)] can help in making native soil

P as well as rock phosphate P move available to crops. It's benefits are seen even with DAP or SSP. Azotobacter and Azospirillum cultures can be usefully utilized under dry land agriculture, where N applications are much less than under irrigation. An all out effort is needed to take these useful research findings on IPNS to the farmers. From the environmental view point IPNS can reduce the nutrient load on soil.

Keywords: *Phosphocompost, FYM, Biofertilizer, Azospirillum, Azotobacter, Biological Properties, Crop-residue, Soil moisture, pH, Press mud cake, Biogas slurry, Azolla, Blue green algae, Rhizobium.*

Integrated Plant Nutrient Supply System

It is an approach, which adopts plant nutrition to a specific farming system and particular yield targets, the resource base, the available plant nutrient source and socio- economic background (41). Since plant nutrients are involved in cycles. PINS involves monitoring all the pathways of plant nutrients in an agricultural production system to maximize the profit on a farm so that farming as a profession is continued, which is the only way to produce food globally (5). Thus, IPNS demands a holistic approach to nutrient management for agricultural production. However, for all practical purposes and in simpler words IPNS involves judicious combined use of fertilizers, bio-fertilizers and organic manures including crop residues and green manures.

IPNS is not new to India and a lot has been written on the subject (7,29,56,57 ,106,112,139,153,157). Krishi Vigyan Kendra, Moradabad a number of On – farm trials were conducted on farmers field and KVK farm. Agriculture being a state subject in India and with a week extension infrastructure IPNS was never put to practice on a large scale. There are several other causes that came in the way of large scale application of IPNS. Firstly, the fertilizer in India is available at a very low subsidized cost and it was doing an excellent job in increasing food production in the country until recently when the secondary (S) and micronutrient (Zn,B, Fe,Cu, Mn *etc.*) deficiencies cropped up in large areas of the country (167) and continuous application of organic manures received little attention.Secondly, making FYM or compost is not a clean operation. It requires dirtying ones hands and the young, educated and neatly dressed up farmers are not willing to do it.Thirdly, green Revolution, which was based on high yielding rice and wheat varieties asked for heavy applications of NPK and it was argued by many agricultural scientist themselves that low nutrient containing farmyard manures and other organic manures are much more expensive than the fertilizer, which also required less labour and was easy to apply. The Farmers were often told that a 50 kg bag of 12:32:16 supplied more nutrients than a tones of FYM or compost. What was ignored was that FYM also supplied secondary and micronutrients and its continued use could help in avoding the deficiencies of these nutrients. Which in a recent yearshave become the key factor in reducing the response of crops to NPK applied through fertilizers. Farmers are themselves realizing that due to declining soil health (42) they have to apply much more fertilizer to obtain the same crop yields which the used to get 10-15 years back. The plant nutrient supply through fertilizer touched

23.01 million tones (mt.) (14.63mt. N, 5.72 mt. P_2O_5 and 2.66 mt. K_2O in 2017 as compared to 1.1mt. (0.74 mt. N, 0.2 mt. P_2O_5, and 0.12 mt K_2O) 1966-67, the pre – Green revolution year. An all out effort has therefore to be made to meet plant nutrient needs from organic sources. IPNS thus assumes great importance and hence the present review, which is in general restricted to the literature generated on the subject during the last25 years or so.

1. Organic Manures

Farmyard Manure/Compost

Farmyard manure (FYM)/compost has been evaluated as asource of plant nutrients in three different ways :- **(i)**its efficiency in comparison to fertilizers. **(ii)** advantage s of its application in addition to NPK (add – on series), and **(iii)** possibilities of economy of N/NPK by FYM (replacement series). Availble information on these aspects it summarized below:

Crop Yield

Relative Efficiency

Not many experiments have been conducted on this aspect in the recent past. In rice- wheat cropping system (RWCS) at Karnal, applicationof 15.30 t FYM/ha to rice was 88-92 per cent as efficient as an application of 150 N 60P and 60 K/ha (150kg N + 60kg P_2O_5 +60kg K_2O/ha; this notation is used through out this paper), while in wheat an application of 20-40 t FYM/ha was only 38-45 per cent as efficientas an application of 150 N, 60P 60K (43). When only FYM was used as asource of plant nutrients there was adecrease of 5-15 per cent in rice grain yield and 35-58 per cent decrease in wheat yield. Higher yield decreasein in wheat could be due to slower decomposition of FYM during cool winter months and there by a reduced plant nutrient release. In maize-linseed cropping system at Bhopal, cattle manure, applied on equal N basis was only 75 per cent as effective as an application of 100 N, 50P, 30K/ha to maize, but equally effective as an application of 50N, 40P,20K/ha to linseed (116). In a 2 year study with sugarcane application of 10 t FYM/ha was 75 per cent as effective as an application of 150 N, 60 P, 60K/ha both in plant ratoon crop (149). Thus when only FYM is applied either on nutrient basis or otherwise it was less effective than fertilizer, its efficiency vaeied withcrop and season.

Add-on Series

In this series of experiments FYM was applied in addition to NPK. A large volume of dataon this series is available from the experiments conducted under AICRP (All India Coordinated Research Project) on long- term fertilizer experiments (162,163). In these experiments yields wre the highest with 100 per cent NPK + FYM. Further, while there was a decline in yield with fertilizeralong specially N, application of FYM along with NPK sustained the productivity of different cropping system (rice-wheat, maize-wheat, soybean-wheat, finger millet – maize *etc.*) at various research centres. For example at Ludhiana NPK +FYM recorded an average increase in grain yield of 17.6 per cent over NPK in maize –wheat system over 29 cropping system cycles (38) (Table 13.1). Increase in grain yield also increases nutrient use efficiency. Application of FYM with NPK increased the recovery efficiency (RE) of N by 45.2 per cent in maize abd 68.2 per cent in wheat. Similarly RE of P increased by 28.4 per cent in maize and 34.2 per cent in wheat and the RE of K increased by 59 per cent in maize and 108 per cent in wheat (32). In a 5 year study at Umian (Meghalaya) application of 10 t FYM/ha along with NPK increased rice yield by 23 per cent (88) (Table 13.4). In a rice-wheat cropping system study at Samastipur (Bihar) application of 10 t FYM/ha along with 80 N, 40 P,20 K/ha to rice and 100 N, 50P, 30K/ha to wheat increased rice yield by 11.9 per cent and wheatyield by 17.9 per cent (33). When 10 tFYM/ha along with 100 per cent NPK was applied to wheat it increased wheat yield by 20 per cent and rice yield by 8.7 per cent (rice received 100 per cent NPK). In a 17 year long –term study on sorghum- soybean/ green gram cropping system at Dharwar (Karnataka) application of 10 t FYM/ha along with 100N,75P,40K/ha to rabi soybe an increased sorghum yield by 16 per cent and soybean/green gram yield by 29 per cent over NPK (66). The advantage of FYM application are many and include supply of plant nutrients and improvement of soil chemical and physical properties (discussed later). It may be mentioned that in addition to macronutrients, FYM also supplies secondary and micronutrients.

Table 13.1. Effect of FYM Fertilization along with or Grain Yield (Maize+Wheat) of the Maize-Wheat Cropping System (MWCS) (Average of 29 cycles), Organic Carbon and Available N, P, K after 29 Cycles of MWCS

Treatment	*Grain*[1]	*Organic C*[2] *(kg/ha)*	*Available N*[2] *(kg/ha)*	*Available P*[2] *(kg/ha)*	*Available K*[2] *(kg/ha)*	*Available Zn*[2] *(kg/ha)*
100 per cent NPK	7.3 (2.56 Maize +4.78 Wheat	682	107	18.4	129	1.49
100 per cent NPKM +10 t FYM to Maize	8.63 (3.52 Maize +5.11 Wheat	1056	148	38.7	162	1.70
C D (5 per cent)		117	9.8	13.7	16.5	0.88

Source: 1: (38), 2: (79).

Economy Series

In this series of experiments part N/NPK is supplied through FYM and the rest through fertilizers. A large amount of data on this series is available from the AICRP on cropping system Modipuram. Hegde (56,57) summarized the results over 15 rice-wheat cropping system cycles at 8 research centers (Palampur, pantnagar, Varanasi, Faizabad, Sabour, Jabalpur, Ranipur and Ranjit Singh Pura) Under this Project. Treatments consisted of economy of 25 or 50 per cent fertilizer N by FYM or green manure in rice or wheat. The best yields were obtained when 50 per cent N needs were met by FYM along with 50 per cent RDF in rice followed by 100 per cent RDF to wheat. This was confirmed by Chauhan *et al.* (35) from Faizabad. This was also true for the rice –rice cropping system at 6 centers (Bhubaneshwar, Chiplim, Jorhat, Rajendranagar, Maruteru and Karamana) (49,99). Data from 4 centers are in Table 13.2. In rice –lathyrus – sesmaumhe cropping system in Nadia (W.bangal) application of 10 t FYM with 50 per cent RDF to rice and sesamum gave 10.5 per cent more rice equivalent as compared to 100 per cent RDF to rice and sesumam (91). In rice-wheat cropping system under rainfed condition at Malan (himachal Pradesh) an application of 5 tFYM/ha along with 50 per cent of RDF (90N,40P,40K/ha) produced 13.7 per cent more wheat as compared to 100 per cent RDF (78). Again under rainfed condition at Giridh (Jharkhand) in rice- based cropping systems applicationof 12 t FYM/ha + 30kgN/ha produced rice yield at par with 60N,30P,30K/ha. but crop yields of succeeding Indian mustard, lentil, barley and linseed were 27-49 per cent more over 30N, 30P, 30K/ha (14). The results suggest that a two crops a year cropping system (rice- wheat, rice- rice and other rice based cropping systems) at leased 25 per cent economy of the recommended dose for the entire cropping can be effected by the FYM.

2. Soil Properties

pH

Most Interesting results of effect of FYM application on soil pH are available from AICRP on long – term fertilizer experiments (LTFE) at Ranchi, where soils are acidic red loams. Continous application of fertilizers (N, NP, NPK) for a period over 40 years reduced the soil pH from 5.4 (control) to 3.9-4.2, while application of FYM along with NPK increased it to 6.3-6.5 (128). Exchangeable Ca increased from 3.05 in control plot to 8.25 Cmol (P+)/kg in FYM added plots. IN this experiment FYM was found as good as lime in improving soil pH.

Organic Carbon

In AICRP – LTFE application of FYM with NPK increased soil organic carbon by 7.7 per cent at Ludhiana, 49.3 per cent at Jabalpur, 33.9 per cent at Bangalore, 14.2 per cent at Ranchi and 50 per cent at Bhubaneswar after 15-25 yearss (162) Similarly in long – term experiments under AICRP on cropping systems, application of 50 per cent N as FYM + 50 per cent of the RDF to rice followed by 100 per cent of the RDF to wheat for 8-9 years resulted in 31 per cent increased in organic carbon at Faizabad and 18 per cent increase in organic carbon at Sabore but there was no significant increase at palampur and Jabalpur (57). IN a two year study on sugarcane, application of 10 tFYM/ha increased soil organic carbon by 40 per cent as compared to an increase of 25 per cent recorded with an application of 150 N, 60P,60K/ha (149). Vineela *et al.* (176) report an increase in soil organic carbon and SBMC due to FYM application in long-term experiments at Anantpur (fellow-groundnut), Solapur (fellow- sorghum), Coimbatore (fingermillet- maize- cowpea), and Barley(maize- chickpea). Small increase in soil organic carbon due to short time application of FYM has been reported by serval workers (33,35,79,104).

Available Nutrients

Application of FYM along with NPK for 27 years at Ranchi increased available N,P,K,S,Cu and Mn (128). Similarly application of 50 per cent N as FYM + 50 per cent RDF to rice followed by 100 per cent RDF to wheat significantly increased available N, P and K over to an application of 100 per cent RDF to rice and wheat at a number of research centres (57). IN a3 year study at new Delhi application of FYM along with NPK significantly increased available N (104) and available Zn, Cu and Fe (89,90). Small increase in availablenutrients in soil due to FYM application have been reported by several workers in short term studies (14,66,110,116).

Soil Moisture

In addition to suppliying plant nutrients, FYM/compost application improves infiltration rate in soil permitting rainwater to penetrate deeper in soil and get stored. This is important for dryland agriculture. For example, incprporation of pigeonpea stalks @ 5 t/ha improved water holdinf capacity of a vertisol and increased the sunflower yield by 44 per cent over to control (119. Ladha *et al.* (73) reported that application of 15 kg N/ha through compost and 10 kg N/ha through fertilizer to maize grown under dryland condition at Arjia (Rajasthan) recorded an infiltration rate of 1.46 cm/hr as compared to 1.17 cm/hr with 25 kg N/ha as fertilizer, the difference was sttisticallysignificant. As result of this average maize yield over 3 crop years out of a total of 6 years of study (3 years had drought and therefore no crop) was 2.37 t/ha with combind application of compost and fertilizer as compared to 1.83 t/ha with fertilizer alone. Again under dryland conditions at Nagpur average seed cotton yield over to 8 Years of study with 90N,19P,37K/ha was only 1.13 t/ha, while that with 5t FYM/ha + 40N,19P,37K/ha was 1.37 t/ha; sorghum grain yields for the same treatment were 2.42 and 2.67 t/ha, respectively (30)

Physical Properties

FYM application inproves soil physical properties. In a study at Umian (Meghalaya) FYM application improved bulkdensity, soilstructure and reduced penetration resistance (88) (Table 13.5). This improvement in soil physical properties resulted in an increase in rice yield by 25 per cent over control.

Biological Properties

A few reports are available on the effect of FYM on biological properties of soils In a long term (2-6yers) study in maize – wheat cropping system at Delhi, application of FYM along with NPK significantly increased soil microbial biomass carbon (SMBC), soil microbial biomass nitrogen (SMBN) and hydrogenase and urease activity over NPK application (75). Ina 2 year study with sugarcane at Lucknow application of 10 t FYM/ha increased SMBC by 21.9 per cent and SMBN by 48.8 per cent over to control as compared to values of 5.6 per cent smbc and 27.9 per cent for SMBN recorded with an application of 150N, 60P, 60K/ha (149). IN a maize-linseed cropping system at Bhopal dehydrogenase activity was significantly more with FYM than with NPK (116). Increase in the bacterial fungal and actinomycetes populations due to FYM application in fingermillet- maize-cowpea cropping system at Coimbatore, maize- chickpea cropping system at Bangalore and in fallow-sorghum cropping system at Sholapur was reported by Vineela *et al.* (176).

3. Legumes in Cropping Systems

Soil fertility restoring capacity of legumes has been known since historic times

(97) even when their capacity to fix atmospheric N was not known. Legumes can fix 50-500 Kf N/ha. depending upon the crop and its growth period. It is estimated that in India legumes fix 2.47 mt. n annually (2). In general forage legumes fix more N. Legumes meet most of their N demands by atmospheric nitrogen fixation with the help of Rhizobia that grow on their root. However, some legumes such as soyabean (Glycine max) are poor starters and need an application of 40-60 Kg N/ha. Rajmash or French been (Phaseolus vulgare) is an extreme example, which when grown in plains need 100-120 kg N/ha., because it has very poor nodulation.

What is more important from the IPNS viewpoint is the amount of N that legumes leave for the succeeding crops. These amounts very from 25-100 kg/ ha. Even when grow as an inter-crop, legumes can spare part of N fixed by them. Banthopadhyay and De (13) reported that when grown in 1:1 row ratio with sorghum, groundnut, mungbean and cowpea contributed 9.6,18.6 and 10.4 kg N/ha to the sorghum respectively. This is why more and more legumes are to be included in the cropping systems. Extensive review on the role of legumes in the cropping system are available (3,48).

Green Manuring (GM)

Green manuring is an age old practice and may be done with non-grain legumes such as sesbania (dhanicha), Crotalaria (Sunnhemp), Stylosanthes, Centrosema and Desmodium or with grain legumes such as mungbean, Urdbean, Cowpea, pigeon pea *etc.* In addition loppings of perennial multi- purpose woody legumes, such as Gliricidia sepium, Cassia siamea and Luecaena leucocephala (Subabul) are widely added to rice fields in south India. Gliricidia a native of tropical central America, was introduced in sout India possibly from Ceylon, while Leucaena was introduced from Australia (64). Joshi mentions that experiments at Konkan Agricultuural University showed that Gliricidia lopping at puddling could produce a crop of 4 tonne/ha. of rice without chemical fertilizer, which is equivalent to about 80 kg/ha. green manures can add 60-120 kg N/ha. and in many situations can meet the entire N demand of a crop to which they are applied. Comparative studies at Pantnagar on nutrient accumulation and rate of decomposition of two Sesbania species (*S. canabina* and *S. rostrata*), Cowpea (*Vigna unguiculation*) and sunnhemp (*Crotalaria juncea*) showed that *S. canabina* accumulated the highest amount of N (97.8 kg/ha at 30 days growth stage and 119.4 kg/ha at 45 days stage. However, for a 60 day crop NPK accumulation was the highest for C. juncea (160.3kg N +20.5kg P +127.2 kg K_2O/ha) (67). In this study C:N ratio in the GM crops was 14-15 at 30 days and 18-19 at 60 days of growth. About 41-43 per cent of the biomass of the 30 day old crops was mineralized in 15 days after incorporation (DAI), while the same amount of biomass from 45 days old crops took at 30 days to decompose. For a 60day old GM crop the decomposition of biomass was 20-30 per cent at 30 DAI. Thus higher the C:N ratio, slower was the decomposition. Considerable research has been done with Sesbania in rice- wheat Cropping system (8,9,68,82,83,84,85,104) and a contribution of 60-80 kg N/ha has beenascribed to it. One of the important finding in recent years has been that a gap of 2-3 weeks between incorporation of GM and rice transplanting as recommended ealier is not required. On the other hand transplanting of rice soon after incorporation of GM leads to better utilization of

N released from GM. Prasad *et al.* (2013) have shown that Sesbania and mungbean residues release N fairly fast, reaching a peak of 30 days after incorporation under submerged conditions, which would coincide with the active tillering stage of rice crop when the N absorption is the fastest. Multilocation trials under AICRP on cropping systems of ICAR have shown that 50 per cent supply of Needed N is possible by green manuring in rice (kharif) in the rice-wheat (56) and rice- rice cropping systems (99). In the same series of trials application of 50 per cent Nthrough GM and 50 per cent NPK to kharif rice and 100 per cent of RDF (Recommended dose of fertilizer NPK) to rabi rice produced 3.5-11.8 per cent more grain than 100 per cent RDF to boththe rice crops, showing that FYM contributes not only N but PK also (49) (Table 13.2). In addition to supplying plant nutrient GM also increases organic C, available N and available K in soil (85,104,134) aggrefates between 0.1 and 0.5 mm size by 62 per cent and reduction in soil bulkdensity due to Sesbania GM in wetland rice. Similary, Sharma etal (133) reported an Improvement in soil structure due to addition on Lantana loppings @ 10-30 t/ha. for 6 years in the rice-wheat cropping systems. In low land of Odisha, growing of horse gram (Dolichos biflorus) during the dry season (November – March) on residual soil moisture and green manuring it before land preparation for rice contributed 43-84 kg N/ha. resulting an economy of N needs of rice by 30 kg/ha (72). Despite N saving, yield increase and improvement in soil physical and chemical properties the area under GM has declined over years in several states (44). For example, in UP it declined from 11lakh ha. in 1994-95 to 0.0 Lha (almost nil 2004-05 (Table 13.3).

Table 13.2. Effect of 50 per cent Contribution of N by FYM/Green Manure/Crop Residue on the Productivity (t/ha) of Rice-Rice Cropping System at Four Research Centers (Results averaged over 15-21 years)

Treatment	*Bhubanesar*	*Maruteru*	*Karjat*	*Karamana*
100 per cent RDF* through fertilizer in khrif and rabi	9.0	10.7	8.2	7.6
50 per cent N through FYM or compost in kharif + 50 per cent RDF in kharif + 100 per cent RDF through fertilizer in rabi	9.8	10.5	7.7	8.1
50 per cent N through crop residue in kharif + 50 per cent RDF in kharif + 100 per cent RDF through fertilizer in rabi	9.4	10.4	7.9	7.9
50 per cent N through greeb manuring/green leaf manure ++ 50 per cent RDF in kharif + 100 per cent RDF through fertilizer in rabi	9.7	10.7	8.5	8.5

*RDF: Recommended dose of fertilizer (NPK).

Source: (49)

The decline during the same period for some other states was Andhra pradesh4.33 to 0.31 Lha, Gujrat from 2.10 to 0.20 Lha, Punjab from 2.59 to 1.43 L ha and Haryana from 0.97 to 0.72 Lha (44). Only Karnataka and Wes Bengal recorded an increase in area under GM. The main reason for decline in area under GM is an increase in area under irrigation, making the farmers to go for remunerative crops.

Table 13.3. Changes in the Area under Green Manure (Lakh hectares) and Rural Compost Produced (Lakh tones) in some States during 1994-95 and 2004-2005 (10 years)

State	*Green Manure*			*Rural Compost*		
	1994-95	*2004-05*	*Change*	*1994-95*	*2004-05*	*Change*
Andhra Pradesh	4.33	0.31	- -	130.34	1.98	- -
Gujarat	2.1	0.20	-	9.30	0.12	- -
Haryana	0.97	0.72	-	74.37	54.18	-
Karnataka	8.74	24.97	++	202.50	587.17	- -
Madhya Pradesh	1.59	1.50	-	24.45	32.00	+
Maharashtra	0.25	0.41	+	16.09	17.83	+
Odisha	2.21	2.23	+	93.18	127.00	+
Punjab	2.59	1.43	-	311.00	322.00	+
Rajasthan	0.59	0.78	+	63.87	63.17	-
Tamil Nadu	0.20	0.96	+	3.71	43.48	+ +
Uttar Pradesh	11.0	0.01	-	950.00	0.04	- -
West Bangal	0.30	1.58	++	275.00	NA	

- Decease, - - Large decrease, + Increase, ++ Large increase.

Source: (44).

The solution to this problem lies in growinh a dual purpose legume (for grain as well as for GM). Considerable research has been conducted on growing mungbean during summer (mid april to mid June) in the rice-wheat cropping system, taking a picking (about 0.5t/ha grain) and intercropping the residue. As regards N contribution and increase in rice and wheat yield such a practice has been found almost equal to Sesbania GM (6,20,111,135,137,144) (Table 13.4) The Improvement in soil fertility with this practice is also of the same order as with Sasbania GM. This practice is gaining popularity in the rice-wheat belt, because the mung bean fetches a good price and provides the farmer money for investing in inputs needed for the rice crop. Also.it adds to the pulse production in the country. Even if 50 per cent of the area under rice- wheat croppings system which is estimated at 10 m ha adopts this practice it can add 2.5 mt to the pulse kitty of the country. Even under rainfed conditions, where farmers mostly take a rabi crop short duration Dual purpose legumes can be grown. For example, Sharma etal (142) showed that incorporation of mung bean and Urd bean residues make available 60-94 kg N and 82-115 kg N/ ha, respectively for the succeeding wheat. GM may be also be sown along with the main crop and incorporated at some stage during the growth period of the main crop. Such a practice may be reffered to as live GM as a corollary to live mulch. For example in direct seeded rice- wheat cropping system at Malan (Himachal Pradesh) (78).

Table 13.4. Effect of Wheat Straw, Legume Residues and Wheat Straw + Legume Residues on the Grain Yield and N Uptake by Rice-Wheat Cropping System (RWCS) and N Balance and Organic Carbon in Soil

Treatment	*Grain*[1] *(t/ha)*	*N Uptake*[2] *(t/ha)*	*N Balance*[2] *(t/ha)*	*Organic Carbon*[2] *(t/ha)*
Wheat straw (WS) (6t/ha)	8.36	175	-1.4	1225
Sesbania green manure (SGM) 5-6(t/ha)	9.00	188	+ 124	1445
SGM +WS	9.70	190	+154	1458
Mung bean residue (MR) (3.5-4 t/ha)	9.25	193	+ 50	1525
MR + WS	9.45	194	+ 99	1568
CD 5 per cent				106

1) Averaged over 2 years of study; 2) After 2 Rice-wheat cycles

Source: (138).

Gm crops namely, cowpea, Sesbania, sunnhemp and soyabean were sown along with rice in the first fortnight of june and incorporated in soil 30-35 days after sowing. When livegreen manured, rice received only 50 per cent RDF. Rice Yields obtained were 78 per cent with sesbania, 81 per cent with sunhemp, 86 per cent with soyabean and 96 per cent with cowpea as compared to that obtained with 100 per cent RDF (90N, 40P and 40K). Thus cowpea live GM can contributed 50 per cent RDF (recommended dose of fertilizer) for rainfed rice. Live GM plots gave as good yield of wheat as obtained with 100 per cent RDF.

4. Crop Residues

Farm animal based traditional Indian agriculture made up a unique time provanbalanced socio- economic ecosystem, where the man took away the grain or other economic produce and the farm animal was contended with the crop residues such ascotton, jute or pigeon pea sticks and part straw/stover was also used for thatching and as kitchen fuel. However, with the mechanization of Indian farms and with the partial urbanization of village where kuccha house are being gradually replaced by piccka or semi-puccka houses, the age old human- farm animal association has been distributed. This has happened quite fast in the north-western rice-wheat cropping system (RWCS) belt, where rice straw become a surplus crop residue in large amounts; about 52.8 million tones (mt.) rice residue is produced from about 10 million ha under RWCS belt (127). Rice residue creates difficulty in its management for **two main reasons: 1)** There is hardly an interval of 2-3 weeks between rice harvested and wheat sowing, **2)** The residue has high moisture content and gernally soiled, so cattle do not like it. On the other hand about 60.8 mt of wheat straw produced annually in RWCS belt is dry and clean and thereis a gap of 8-10 weeks between wheat harvested and rice transplanting, so it is possible to transport it to the bordering states of Rajsthan and Madhya Pradesh, where it fetches good money. Most of thereview on residue management therefore is restricted to rice and wheat residue, which is about one-fourth of the total crop residues producedin the country and also where bulk of the research on the subject has been conducted. Sankar (127) estimated that out of the total rice and wheat residue of 113.6 mt about one-third (37.87 mt) is available for recycling, the associated nutrient content being 0.634 mt of N+P+K. One tone of rice residue contains 6.1 kg N, 0.8 kg P, and 11.4 kg K, while one tone of wheat residue contains 4.8 kg N, 0.7 kg P and 9.8 kg K. Nearly three- fourth of the area under RWCS is being combine harvested, which leaves stubbles on the field and straw spread on the field is difficult to gether. The most common practice with the farmers is to burn the rice residue in situ, which lead to the loss of plant nutrients and precious organic carbon. It may be mentioned that Indian soil are already poor in organic carbon (generally 0.3 to 0.5 per cent in the RWCS belt). Further burning of rice residue also produces CO2 and other green house gases (GHG). About 26 mt. of CO_2 0.09 mt. Ch2 and 0.28 mt N2O is produced annually in the RWCs belt in Panjab, Harayan and Uttar Pradesh due to rice residue burning (126). Also residue burning produses smoh wich is hazardus to health specilly to persons suffering from "Asthma. No data on this are available. As regards nutrient loses, Indian data are not yet available but Kumar *et al.* (68) from New Zealand reported that for every tone of wheat straw burned 2.4 kg N was lost. In Australian study (74) S loss from the burning of rice straw varied from 40-60 per cent. Heard *et al.* (54) from Canada reported that burning of wheat, Oat and Flex straw led to 98-100 per cent loss of N and 75 per cent loss of S.

There have been a number of studies on the effect of incorporation of rice or wheat or both residues in the RWCS (12,34,107,111,115,129,130,140,141,). In some of these studies an increase in grain yield was obtained, while in others grain yield remained unaffected. Hower, it is clear that crop residues can be safelyincorporated in the RWCS without any detrimental effect s. Sharma and Prasad (135) also studies the effect of inoculation of rice and wheat residues with **celluloytic fungal** culture of *Trichrus spiralis* for hastening the decomposition of crop residues and on the grain and straw yield of rice. Inoculation with **Trichrus spiralis** showed no advantage. They, however, reported a significant increase (18 per cent) in K uptake by RWCS when residue were incorporated. When wheat is sown after rice straw incorporation in RWCS some set back in the initial growth stage of wheat has been reported due to temporary immorilisation of native soil or applied N (113) and some workers recommended an additional starter N application of 25 kg N/ha at wheat sowing to overcome this problem. In the case of rice after wheat an alternative to additional starter N application is mixed application of wheat residue and sesbania or mung bean residue (8,134). Sharma and Prasad (138) reported an increase in the rice + wheat grain yield of 7.7 per cent, 16.1 per cent and 25.2 per cent with rice/ wheatstraw, sesbania GM and sesbania GM + rice/wheat straw, respectively; the values of rice/wheat straw, mung bean GM and mung bean Gm + rice enriched straw were 7.7 per cent, 19.3 per cent and 21.9 per cent respectively. Surekha *et al.* (159) also reported an increase of 13.6 per cent, 17.9 per cent and 27.2 per cent due to incorporation to incorporation of 5 t/ha of rice straw, 10t/ha of sesbania GM and 2.5t/ha of rice straw + 5t/ha of sesbania GM, respectively. The increase due to 10/ha of mungh bean GM and 2.5t/ha of rice straw +5t/ha of mung bean GM was 28.5 and 27.5 respectively. Thus if possible rice or wheat residue may be applied in mixture with some legume residue for better results. In rice-wheat cropping system application of 50 per cent N through crop residue and 50 per cent RDF to kharif rice followed by 100 per cent NPK to rabi rice was at par with 100 per cent NPK to both kharif and rabi rice (99). In these trials incorporation of rice residues was at par with FYM at all centers but inferior to GM at 2 out of 5 centers. In rainfed rice in Meghalaya, incorporation of forest grass or Ambrosia residue resulted in a 19-20 per cent increase in rice yield (88) (Table 13.5). In sugarcane about 10-20 per cent of the weight of cane harvested is left as trash on the field (185), which contains about 0.4 per cent N, 0.13 per cent P and 0.5 per cent K (149). The trash is generally burnt by the farmers and a ratoon crop is raised. Studies at Lucknow (184) have shown that leaving the trash as mulch increased the yield third ratoon by 28 per cent over that obtained when the trash was burnt. In another study at the sameresearch center incorporation of 10 t/ha trash along with culture of Trichoderma for hastening its decomposition increased the yield of plant cane by 26.9 per cent and that of first ratoon by 32.4 per cent. The cane yield obtained with the incorporation of trash + Trichoderma was 809 per cent of that obtained with RDF (150N, 60 P and 60K) both in plant cane and ratoon (149). In cotton, incorporation of cotton stalks, wheat residue and cotton stalk + wheat residue increased seed cotton yield by 4 per cent and 14 per cent, respectively (21). Thus incorporation of cotton stalk was less effective due to its slower decomposition. Babala and Itnal (10) also reported that incorporation of crop residues in cotton + soybean intercropping increased the productivity of the system. They also reported benefits of inoculation of crop residues with celluloytic fungus phaneorochaete crysoporium in hastening the decomposition of residues.

Table 13.5. Effect of Grass Residues and FYM on the Grain Yield of Rice (Average over 5 years) and some Physical Properties of Soil at Umian (Meghalaya)

	Rice Grain (t/ha)	*Bulck Density (mg/m³)*		*Peneteration Diameter (MPa)*		*Mean Weight Diameter (mm)*	
		0-15 cm	*15-30 cm*	*0-15 cm*	*15-30 cm*	*0-15 cm*	*15-30 cm*
No residue	2.89	1.26	1.25	2.21	2.00	2.00	2.17
Grass-1 @ 10 t/ha	3.46	1.05	1.12	2.62	2.14	1.70	1.82
Ambrosia @ 10 t/ha	3.44	1.05	1.10	2.48	2.21	1.71	1.82
FYM @ 10 t/ha	3.56	0.91	1.04	2.77	2.35	1.54	1.63
CD at 5 per cent		0.01	0.13	0.19	0.13	0.11	0.20

1. Mixture of goat grass (*Ageratum oncyzoides* L), quick weed (*Galinsoza parvilora* Cav.), *Eupatorium odoratum* L.), beggartricks (*Bidens pilosa* L.) and thatchgrass (*Imperata cylindrical* Beauv.) containing 1.44 N, 0.15 per cent P and 0.86 per cent K (on dry matter basis).
2. *Ambrosia artmissifolia* L. Containing 2.15 per cent N, 0.28 per cent P and 1.64 per cent K (on dry matter basis).

Note: Rice received 80 N +60 P +40 K. Soil pH 5.1, organic carbon 1.92 per cent available P 10.3Kg and available K 112.3 kg/ha.

Source: (88).

In rice wheat cropping system most researchers (12,17,111,132,134,135,183) have reported an increase in soil organic carbon and total N when crop residues were incorporated. In sugarcane also leaving of trash as mulch increased soil organic carbon by about 25 per cent, available N by 6 per cent and available P by 10 per cent over the burning of trash (184). In corporation of trash with Trichderma increased soil organic carbon by 40.6 per cent, total N by 34.2 per cent, SMBC by 11.0 per cent and SMBN by 42.2 per cent (149).

Incorporation of crop residues in RWCS also improve soil physical properties, suchas water stable aggregates, bulckdensity and penetration resistance (83,84,88) (Table 13.2). Incorporation of crop residues in RWCS also increase bacterial and fungal population (144). This improvement in soil health is a major grain in incorporation of rice/wheat residues in RWCS and this is important from the sustainability view point. Incorporation of part/full rice residue has come to stay due to the development of direct seeding drills and this practice is gaining ground in the RWCS belt in northern India (52). Under dry land condition crop residues could be used as amulch. As amatter of fact this how crop residue management and zero- tillage started in the mid –west of USA (105). A four year study study at dry land research centre, Bellary, showed that whenDolichos was grown in between sorghum rows and cut at 45 days after sowing and left in situ as mulch it increased sorghum yield by 44.4 per cent (93). This practice also reduced run off water loss by 9.1 per cent and soil loss due to erosion by 12.2 per cent. Ita alo increased soil organic carbo

5. Other Organic Manures

Vermicompost (VC)

Compost made with the help of earthworms is known as vermicompost. The general analysis is 1.9 per cent N, 2.0 per cent P and 0.8 per cent K,100mg/kg Zn, 40mg/kg Cu and 500mg/kg Mn (148). The exoyic species used for making VC are *Eisena foetida, Eudrilus euginiae* and *Perionyx excavates,* while indigenous species include Lampito mauriti, Octachaetonaserrata and Perionyx sansibaricus (155). VC gained popularity due to recent intrest in organic farming (109) and is mostly used in vegetative and fruit production. There have been only a few agronomic studies on VC. IN wet season rice at Sriniketan (W.B.), application of VC (1.6 per cent N, 1 per cent P, 1 per cent K) @ 10 t/ha along with 50 per cent RDF was superior to 100 per cent RDF (60N, 30P,30K) and produced 16 per cent more rice, while an application of FYM (0.5 per cent N,.2 per cent P, 0.5 per cent K) @ 10 t/ha along with 50 per cent RDF was at par with 100 per cent RDF (15). These results show superiority of VC over to FYM, which could be partly due to its higher NPK content. Similar advantage of VC over FYM was reported by Bhattacharjee *et al.* (22). In direct seeded upland rice at Malan (Himachal Pradesh) an application of 2.5 t VC/ha produced only RDF (80N,40P,40K/ha) while an application of 50 per cent RDF and 1.2 t VC/ha produced 78 per cent grain yield of that obtained with RDF (77). In sugarcane, however, an application of 10 t VC/ha produced same yield of plant and ratoon crop as an application of RDF (150N,60P, 60K) (149). Application of VC recorded significantly more organic C, total N, SMBC and SMBN than RDF. In maize – linseed cropping system at Bhopal, when applied on equivalent N basis (8.4 t/ha), VC was only 79.5 per cent as effective as NPK (100N,50P, 30K/ha) in maize but 107.7 per cent as effective as NPK(50N,40P,20K/ha) in linseed (116).

Poultry Manure (PM)

Poultry manure is richer in P than other organic manures. The general analysis is 1.8 per cent N, 2.5 per cent P and 1.4 per cent K and the release of plant nutrients from PM is fairly fast. IN plant a laboratory study 45 per cent of total N in PM was mineralized in 4 weeks as compared to only 12 per cent in FYM (189). IN a field study at Ludhiana 4 tPM + 60kg N/ha was as efficient as 120kg N/ha as urea (186) for rice. Further application of PM in rice had significant residual effects in succeeding wheat. Bijai singh *et al.* (23) also showed that Pm –N was as efficient as urea-N in increasing yield and N uptake in rice. However, when applied on equal N basis in maize- linseed cropping system Pm was only 82.5 per cent as effective as 100N,50P, 30K/ha in maize but 123.6 per cent as effective as 50N,40P,20K/ha in linseed (116). At the end of 2 cycles of maize –linseed cropping cycle PM recorded 0.68 per cent organic C in soil as compared to 0.5 per cent for NPK and hydrogenease ctivity was also significantly more with PM than with NPK.

Biogas Slurry (BGS)

Biogas sluary is the residue left in biogas units after the fermentation and liberation of biogas. Anaerobic fermentation of cattle dung in biogas units increases its nutrient content. The gernal analysis is 1.4 per cent N,1.2 per cent P and 1.0 per cent K. Experiments conducted at IARI (New Delhi) showed htat N applied through BGS was as effective as urea for rice and wheat (59). At Ludhiana application of 40 per cent N through BGS and the rest 60 per cent through ures was as efficient as 100 per cent through urea (98). In another study at Ludhiana, application of 60 kg N/ ha as BGS and 60kg N/ha as urea produced more rice and wheat than 120kgN/ha as urea (51). In a rice-wheat cropping system (RWCS) study at IARI, when applied on equal N basis BGS was 10.7 per cent were efficient than 100 per cent N as urea, while 70 per cent N through urea and 30 per cent N through BGS was at par with 100 per cent N through urea (19). The highest grain yield of RWCS was obtained with 70 per cent N through pusa neem micro emulsion coated urea +30 per cent N as BGS, which was 17.5 per cent more than100 per cent asurea.

Pressmud Cake (PMC)

PMC is a waste product of sugar industry and about 8.9 mt is produced annually in India. The general analysis is 1.6 per cent N, 1.0 per cent P and 0.8 per cent K. IN a study at Ludhiana application of 5 tPMC/ha along with 40-60 kg N/ha as urea produced rice yield equal to that obtained with 120kgN/ha as urea(187). Further residual effect on succeeding wheat was euivalanet to 40 kg N and 13 kg P/ha. At faizabad, application of 50 per cent N through PMC and 50 per cent through urea to wheat was at par with 120kg 60P and 40K/ha (35). Thus supplying N through PMC also met the P and K needs of the crop. Application of PMC recorded significantly more organic C and available N,P and K. IN sugarcane, application of 10 t PMC/ ha was at par with an application of 150N 60P 60K/ha (149). Application of PMC also increased soil organic C by 50 per cent and total N by 60 per cent over control. It also improved soil biological properties and there was a90.8 per cent increase in SMBN over control.

Bio-compost (BC)

Bc is prepared by mixing PMC with spent wash from the distillery units (1). The general analysis is 1.9 per cent N, 1.85 per cent P, 1.48 per cent K,25.5mg/kg Zn, 34.8mg/kg mn, 9.1mg/kg Cu, and 5.9mg/kg Fe. In astudy at New Delhi application of 5 t BC/ha along with 50 per cent RDF produced the highest yield of wheat and maize (21.8 per cent more of wheat and 6.6 per cent more of maize cobs) than 100 per cent RDF (120N 60P 60K) (169). Application of BC with NPK also resulted in a significant increase in soil organic and available N,P,K,Zn,Fe,Cu and Mn. Kavimani *et al.* (65) reported that an application of 2.5 t BC/ha along with NPK recorded higher yields of pods in groundnut than NPK alone.

Phospho Compost (PC or PEC)

Phospho compost or phosphate enriched compost is the compost enriched with P. The P enrichment may be done with single superphosphate (SSP) or ground rock phosphate (RP). When SSP is used some loss of NH3 during composting can be reduced making compost richer in N aswell as P. Pc has been mostly made with RP specially low- grade RP such as Mussoorie rock phosphate. In a composting study at Parbhani, Maharashtra (39) using NADEP technique for compost making total P content in compost increased from 0.69 to 0.92 per cent with RP, to 0.98 per cent with SSP and to 1.03 per cent with RP + PSB (*Aspergillus awamori*) (Table 13.6). Composting RP with PSB also recorded the highest citrate soluble P. Increase in total and citrate soluble P due to RP enrichment of compost was ealier reported by several workers (81,86,87). This makes RP-P readily available even on calcareous soils.

Table 13.6. Effect on Total and Citrate Solution P in Compost after 150 Days of Composting Sorghum Stubble along with RP, SSP or RP + PSB

Treatment	*Total P (Per cent)*	*Citrate Soluble P (Per cent)*
RP + PSB	1.03	0.59
SSP	0.98	0.52
RP	0.92	0.49
Control	0.69	0.31
CD per cent	0.032	0.019

Phospho Gypsum (PG)

Phospho gypsum is a by product of phosphate fertilizer industry and about 608 mt PG is produced annually in India. As a contrast to organic manures, which are applied for NPK, PG is a rich source of S and Ca. It contains 16 per cent S and 21 per cent Ca (28) and assumes great importance in view of wide spread S deficiency in the country (27,108,164,166). PG has an advantage over elemental and organic S sources, in that the letter must undergo microbial oxidation before becoming available to crop plant, while S from PG is readily available immediately after application. In an incubation study on an acid soil ultisol (pH 4.1) at Vallayani (Kerla) available S 12 days after incubation in PG treated soil was 29mg/kg as compared to only 13 mg/kg soil in control (80). Exchangeable Ca 12 days after incubation was 2.56 C mol/kg in PG treated soil and 0.98 C mol/kg in control. Shahu and Nayak (125) repeated from Bhubaneswar that PG was as effective as SSp as asource of S for groundnut and significantly superior to pyrites for mustard ab increase in yield of 26 per cent in groundnut, 102 per cent in sesmum, 23 per cent in green gram and 7 per cent in tomato with PG + RDF over RDF.

Sewage Sludge

Sewage sludge consists of water and bio soilds carried from the residential buildings and industrial units and surface and ground waters that enter the sewage system. Sewage biosolids contain 2.4-6.7 per cent N, 2.6-3.1 per cent P, 0.3-0.9 per cent K, 1.2-7.5 per cent Ca, 0.5-0.8 per cent Mg and 0.2 per cent sulphate- S. Although good yields of serval crops with sewage study have been reported (61), it remains mainly in the domain of pri-urban agriculture and mostly for vegetable and some fruit crops. Use of sewage sludge for crop production is away of disposing it off. Althrough sewage sludge contains about 33 per cent organic matter and helps in building up soil fertility, it contains pathogenic bacteria, viruses and protozoa along with some prarsitic helminthes, which can be hazardous to human health (181. The other drawback is its high content of heavey toxic metals such as Pb, Cd,Cr, Co and Ni (156). Sewage has therefore to be properly ttreated before its application to farm fields (131). Use of sewage sludge even in peri-urban agriculture has to be done only with due care.

Bio-fertilizers

Strictly speaking the term bio-fertilizers is amisnomer, since as such the bio-fertilizers do not supply plant nutrients. Biofertilizers are cultures of mircorganizers that are capable of fixing atmospheric N or solubilising rock phosphate P and mobilizing native soil P or solubilising native soil K (Table 13.7). Bio-fertilizer cultures are generally used for inoculating seeds or other planting materials. The term biofertilizers is also used for cultures of microganisms used for accelerating the decomposition of organic meterials in composting or directly in fields with crop residues. Some biofertilizers and Azospirillum have been used for enriched of compost.

Rhizobium

Rhizobium inoculants have been the most successful biofertilizers in India. The first Rhizobium inoculants was produced by M.R. Madhok, while the commercial production of Rhizobium cultures was started in 1956 (165).

Rhizobium are specific to crop species and proper nodulation and maximization of N- fixation appropriate culture is amust. For example, for field beans, lentil and lathyrus, a culture of R. leguminosarum is right choice, while for chickpea a culture of Mesorhizum ciceri is required and for mungbean, urdbean, cowpea and

Table 13.7. Kinds of Biofertilizers

A. Nutient fixers/Mobiliseres/Solubilisers
A.1 Nitrogen fixers
A.1.1 Symbiotic - Rizobium, Azolla – Anabaena
A.1.2 Non symbiotic Associate: Azosprillium Acetobacter Non-associative: Azotobactore, BGA (blue green algae).
A.2 Phosphate mobiliser/solubilisers
A.2.1 Phosphate mobiliser – VAM (Vesicular arbuscular mycorhyza)
A.2.2 Phosphate solubilisers Bacteria : Bacillus megaterium, B. polymyxa, B. subtitls, Pseudomonas striata, P. rathoris. Fungi : Aspergillus awamori, A. niger, Penicillium digitatum Actinomycetes : Schwanniomyces occidentalis
A.3 Potassium mobilize Frateuria aurentha
B. Biodecomposition accelerators
B.1 Cellulolytic - trichoderma, Celumonas, Myrothesium, Paecillimyces B.2 Lignolytic - Arthobacter, Humicola

Source: (165).

groundnut a culture of Rhizobium sp. is desirable. In India the greatest success has been achieved with R. japonicum in soybean mainly because is a poor nodulator and unless inoculated with Rhizobium its growth and yield remains poor. The population of Rhizobium is generally low in tropical soils to high tempetature and moisture stress. As regards moisture stress venkateswarlu and Rao (175) reported that Rhizobium population and nitrogen fixation declined drastically in mungbean, mothbean and clusterbean as the soil moisture potential crossed - 0.2 M Pa. The deleterious effect of temperature are so heavy in the soybean belt of Madhya Pradesh that even revial after the onset of monsoon gave a population of only 436-1410 cell/g soil (117). In a survey of chickpea growing fields Gwalior, about 40 per cent field had <100 Rhizobium/g soil (124). Similarly poor nodulation (10-20 nodules/ plant) was recorded in 60-100 per cent surveyed area under pigeon pea, blck gram and lentil grown in 43-47 villages in each of the three district by Madhya Pradesh (94). Similarly in a survey in Ranipur, Durg, Rajendgaon, Bilaspur and Chhatisgarh district of Chhatisgarh 68 per cent soil samples did not have native Rhizobia of one or more than one legume crop (53). Rhizobium inoculation of pluses can thus greatly increase N- fixation and grain yield.

Azotobacter

Azotobacter, a nonsymobotic N fixer has been successfully used in Sovit agriculture and was indroduced from there, Azotobacter inoculation leadsto a fixation of 10-25 K N/ha (100). It is therefore of significant in dryland agriculture, where crops are generally less fertlised than under irrigated conditions. A number of early experiments demonstrated the beneficial effects of Azotobacter inoculation on maize, sugarcane, rice, tomato, onion and mustard (101). In later experiments yield benefits in sorghum, pearl millet and rainfed cotton with Azotobacter inoculation were 15-20 per cent,.0275 and 11-16 per cent, respectively. Studies conducted at the Central Institute of Cotton Research, Nagpur, showed that Azotobacter inoculation with 40 Kg N/ha produced seed cotton yield equal to that obtained with 60 kg N/ha (18). Trials on cotton in Vidarbha region under the national Agricultural Technology Project (96) also showed the advantage of Azotobacter inoculation. However, Prasad and Prasad (103) at Delhi failed to obtain any advantage of Azotobacter inoculation in cotton. In mustard Chauhan *et al.* (34) reported an yield increase due to Azotobacter inoculation equivalent to 25-30 kg N/ha. Under irrigated conditions Kumar and Ahlawat (69) reported an increase in wheat yield due to Azotobacter inoculation equvilant to 15-20 kgN/ha.

Azospirillum

Dobreiner and Day (40) from Brazil were the first to report – fixation of N by spirillum lipoperum on the roots of grass Digiteria decumbens. Azospirillum just like Azotobacter fixes about 10-25 kg N/ha. Multi-location trials under AICRIP on millets showed that Azospiriluum brasiliense inoculation increased the grain vield of pearl of millet by 17 per cent at 6 and that of sorghum by 7-13 per cent at 4 out of 9 locations (158). Chauhan *et al.* (34) reported an yield increase in mustard due to Azospirillum inoculation equivalent to 25-20 kg N/ha, they found Azospillium to be slightly better than Azotobacter. Azospirillum inoculation significantly increased the pod yield in groundnut (12). Reddy and Sudhakarababu (119) observed that Azospirillum inoculation ontributed N needs of sunflower (*Heliauthus annus* L.) by 50 per cent. However, Venkateswarlu (173) from a review concluded that the results with Azospirillum inoculation under dryland conditions weer not consistent. In addition to N fixation both Azospirillum and Azotobacter improve plant growth through the production of phytohormones leading to better root growth (174,177,178).

Blue Green Algae (BGA)

Research on BGA as N- fixer in rice fields was first initiated at Banaras Hindu University (152) and was followed up at IARI (161,171) and Central Research Institute (CRRI) (150,151). BGA can fix 25-30 kg N/ha in a rice season. Reviewinh the data from field experiments from different countries Roger and Kulassoriya (121) observed that BGA inoculation can increase the rice yield by 14 per cent. Singh *et al.* (147,154) observed that BGA inoculation is complementary to green manuring with sesbania and use of neem cake coated urea (NCU) in rice. BGA inoculation was more effective with neem cake coated urea than uncoated urea. BGA inoculation with 60 kg NCU gave grain yield of rice which did not significantly differ than that obtained with 120Kg N/ha as prilled urea.

Azolla

Azolla has been used for N fixation in rice in China since 6th century (36). Research on Azolla in India was initiated at Central Rice Research Institute, Cuttak

(Odisha). Azolla does not fix N per se but has an Algae Anabana associated with it whichfixes the N. Azolla prefers low temperatures (16-14 0C) and thus performs best under low temperature conditions but cultures have been identified that can survive upto 30 0C. Under field field conditions it fix 30-40 kg N/ha but needs 10-25 kg P_2O_5/ha for doing this job. It therefore is a difficult choice for India, since it imports most of its P either as phosphate rock or DAP. Singh *et al.* (154) reported that inoculation with Azolla recorded a significant increase in rice yield in plots receiving no nitrogrn (control) as well as in plots receiving 60 or 120 kg N/ha. They also recorted a significant increase with the combined application of BGA and Azolla, suggesting that the two inoculations are compatiable.

Pulses

In chickpea an increase of 14-30 per cent due to Rhizobium inoculation was reported from central and western India (26). Data from AICRP on pigeonpea suggest an increase of 13-52 per cent in yield due to Rhizobium inoculation (4). Results on inoculation with Bradyrhizobium in pluses crops at research centres under dryland research project of ICAR showed an increase in yield of 2.9-14.5 per cent in mungbean, 6.4-14.2 per cent in clusterbean, 6.2-38 per cent in pigeonpea, 8.2-22 per cent in chickpea and 5.0 to 6.5 per cent in soybean (172). Since soil misture affects nodulation by Rhizobium, the moisture conservation practices under driland conditions can affect it. Pigeonpea, soybean and mungbean showed higher nodulation when planted on raised bed with conservation furrows as compared to flat beds, however, in chickpea flat beds produced more titlers (179). Venkateswarlu (171) reported higher nodulation, nitrogenase activity abd seed yield in mungbean and pigeonpea withconservation furrows as compared to flatbed. Gupta and Rao (50) reported that mulching with weed biomass provided a more favourable temperature regime and increased nodulation, nitrogenase activity and grain yield in cowpea and mungbean. Depth of sowing also infiuence nitrogen fixation specially in groundnut where considerable nodules are found on hypocotyle. Deep sowing alos reduced nodulation in chickpea in heavy black soils (123). Some reports are available on the benefit of dual inoculation with Rhizobium and PSO (Phosphate solubilising organisms). Jain *et al.* (62) reported that dual inoculation produced the highest grain yield in chickpea. In another study (120) dual inoculation in chickpea gave yield equivalent to an application of 18 N 46P/ha. Similarly in pigeonpea, dual inoculation with 75 per cent RDF gave the same yield as with 100 per cent RDF (94). In mungbean also dual inoculation with 40kg P_2O_5/ha showed an advantage in seed yield (146).

Leguminous Oilseeds

Averaged over 29 trials, the increase in yield of groundnut due to Rhizobium inoculation was 272 kg/ha, about 10.3 per cent over un- inoculated control (60). Wankhede *et al.* (180) also reported an increase in groundnut yield due to Rhizobium inoculation. An increase in soybean yield due to Rhizobium inoculation has been reported by serval researchers (25, 37,58). Application of molybdenum (180) and pellerting with calcium (102) helps in making Rhizobium inoculation more effective.

Phoaphate Solublising Organisms (PSO/PSB/PSF)

Phoaphate Solublising Organisms have received considerable attension in recent year (46,47). Release of organic acids (gluconic, lactic, citric, tartaric *etc.*) by PSO has been suggested as the main cource for their phosphate solubilising ability (11). Application of pso along with ground rock phosphate is reported to increase the grain yield in rice (92), wheat (16,160,168), sorghum (63), maize(16), groundnut (55,76,118) and sunflower (114). Sharma and Prasad (137) reported that in rice- wheat cropping system application of Mussoorie rock phosphate (MRP) along with PSB was at par with DAP. Further, application of rice and wheat residues along with PSB and MRP improved the efficiency of PSB (Table 13. 8). Gaur (46) has also observed that application of organic matter improves the efficiency of PSO. PSO's have been widly evaluated with ground rock phosphate (RP) for pulses and data are available for a number of trials conducted under AICP on chickpea and MULLaRP during 1996-2000 in different parts of the country (45).

Table 13.8. Effect of Rice Residue and PSB on the Productivity and P Uptake in Rice-Wheat Cropping System (Average of 3 cycles) and available P at the End of 3 Cycles of RWCS

Treatment	*DAP*	*MRP*	*MRP+PSB*
Grain Yield of rice + wheat (t/ha)			
No residue	8.4	7.5	8.2
6 t/ha of rice straw	9.2	8.4	9.2
P uptake by rice + wheat (kg/ha)			
No residue	25.6	22.6	25.1
6 t/ha of rice straw	28.5	25.7	28.4
No residue	13.2	22.9	26.8
6 t/ha of rice straw	24.0	33.8	34.8

Source: (137).

On average, application of 40 kg P_2O_5/ha as RP increased chickpea yield by 9.5 per cent over no P content as compared to 21.8 per cent obtained with DAP. PSO's increased chickpea seed yield both with RP and DAP and PSB (phosphate solubiling bacteria) perefomed better (10 per cent increase) than PSF (phosphate Solubiling fungi) 6-7.8 per cent increase). In mungbean the increase in grain yield due to PSB or PSF inoculation with 20 kg P_2O_5/ha as DAP or RP was about 5-6 per cent.

Thus the benefits of PSO inoculation are not restricted only with difficult soluble P fertilizers such as RP bit are also obtained with DAP.

Conclusion

In India where all of it's potash and bulk of its phosphorous requirements have to bemet by imports, there is no alternative to IPNS, specially in the face of recent spiral rise in fertilizer prices in the international market. By 2025 India's plant nutrient needs are estimated at 45million tones. Available research suggests that 20-25 per cent plant nutrient needs can be met by recycling from and industrial waste and by the use of biofertilizers. Continued use of these organic materials will also help in partly overcoming secondary and micronutrient deficiencies, which have cropped up all over the country. Amounts of plant nutrients added through rain and irrigation waters,soil deposits due to erosion in low lands from the upper resions and silt deposits due to floods also need to be measured. IPNS is the only way to make Indian agriculture sustainable and in future for better soil health.

REFERENCES

Agrawal, P.K. and Agrawal, S. (2003). Proc. 65th Ann. Conv. Sugar Tech. Assoc. India, Bhubaneswar. August 22-24.

Ahlawat, I.P.S. and Gangriah, B. (2004). Souvenir National Sym. Resource Conserv. And Agric. Productivity, Ludhiana, Nov. 22-25, p. 32-36.

Ali, M. (2003). In Assessing the Resource Conservation Issues in Rice-Wheat Systems in South Asia-A resource book, Rice-Wheat Consortium for IGP – CiMMYT Centre, New Delhi, p. 210-214.

Ali, M., Singh, K.K. and Saad (2004). *Fert. News,* 49(4): 43-46.

Ange, A.L., (1997). Proc. Nat. Workshop IPNS, March 10-12, Bhubaneswar, IFFCO, New Delhi, p. 16-33.

Antil, R.S., Singh, B., Kumar, V. and Singh, M. (1989). *Indian J. Agron.* 34: 213-216.

Aulakh, M.S. and Grant, C.A. (2008). *Integrated Nutrient Management for Sustainable Crop Production*, Howorth Press, USA.

Aulakh, M.S., Khera, T.S., Doran, J.W. and Bronson, K.W. (2001). *Soil Sci. Am. J.,* 65: 820-827.

Aulakh, M.S., Khera, T.S., Doran, J.W., Singh, Kuldeep and Singh, Brijai (2000). *Soil Sci. Soc. Am. J.,* 55: 1020-1025.

Babala, H.B. and Intal, C.J. (2004). Proc. Int. Symp. Start. Sustain. Cotton Prod.- AGlobal Vision, USA, Dharwad Vol. 2: 92-97.

Bagyaraj, D.J., Krishnaraj, P.U. and Khanuja, S.P.S. (2000). Proc. Indian Nat. Sci. Acad. B66, 60-82.

Balasubramanian, P. and Palaniappan, S.P. (1994). *Indian J. Agron.,* 39: 131-132.

Bandopadhyay, S.K. and De, R.J. (1986). *Agric. Sci. Camb.,* 107: 629-663.

Banik, P. and Sharma, R.C. (2008). *Indian J. Agric. Sci.,* 78: 240-243.

Barik, A.K., Das, A., Giri, A.K. and Chattopadhyay, G.N. (2006). *Indian J. Agric.Sci.,* 76: 657-666.

Benarjee, M., Rai, R.K., Maiti, D. and Dhar, Slivia (2006). *Indian J. Agric. Sci.,* 76: 753-755.

Beri, V., Sidhu, B.S., Bahal, G.S. and Bhat, A.K. (1995). *Soil Use Manage.,* 11: 51-54.

Bhale, N.L., Sahay, R.K. and Narayan, S.S. eds (1985). CICR- First Decade 1976-85, Central Inst. Cotton Res., Nagpur, p. 23-24.

Bharade, N.M., Shivay, Y.S., Singh, S. and Kumar, D. (2006). *Indian J. Agric. Sci.*, 76: 550-553.

Bhargavi, K., Raghva Reddy, C., Yellammanda Reddy, T. and Srinivasulu Reddy, D. (2008). *Indian J. Agric. Sci.*, 78: 170-172.

Bhaskar, K.S., Singh, J., Blaise, D., Wasnik, S.M., Majumdar, G. Tuteja, O.P., Dongiwal, G. and Mendha, P.N. (2002). Extended Summarise, 2nd *Int. Agron. Congr.* Vol. 2: 804-805.

Bhattacharjee, G., Chaudhari, P.S. and Dutta, M. (2001). *Asian J. Mircob. Biotech. Environ.*, 3: 191-196.

Bijai Singh, Yadvinder Singh, Maskina, M.S. and Meelu, O.P. (1997). *Nutr. Cycl. Agroecosyst.*, 47: 243-250.

Bijai Singh, Bronson, K.F., Khera, T.K. and Pasquin (2001). *Natr. Cycl. Agroecsyst.*, 59: 227-237.

Bist, J.K. and Chandel, A.S. (1994). *Indian J. Agron.*, 39: 129-132.

Biswas, B.C., Das, S. and Kalwe, S.P. (2001). *Fert. News*, 46(2): 15-28.

Biswas, B.C., Sarkar, M.C., Tanwar, S.P.S. and Kalwe, S.P. (2004). *Fert. News*, 49(10): 13-33.

Biswas, R.P. and Sharma, P.D. (2008). *Indian J. Fert.*, 4(8): 49-52.

Blaise, D. and Prasad, R. (2005). *Indian J. Fert.*, 1(1): 37-44.

Blaise, D., Singh, J.V., Venugopalan, V. and Mayee, C.D. (2003). *Acta Agron. Hung.*, 51: 61-67.

Boparai, B.S., Yadvinder Singh and Sharma, B.D. (1992). *Arid Soil Res. Rehab.*, 6: 135-143.

Brar, B.S. and Pasricha, N.S. (1998). In: Long-term Soil Fertility Management through Interated Plant Nutrient Supply (A. Swarup *et al.*, eds.) Indian Institute of Soil Science, Bhopal, p. 154-158.

Chaudhary, S.K. and Thakur, R.B. (2007). *Indian J. Agric. Sci.*, 77: 443-444.

Chauhan, D.R., Paroda, S. and Singh, D.P. (1995). *Indian J. Agron.*, 40: 639-642.

Chauhan, R.P.S., Kumar, R., Singh, O.P. and Prakash, V. (2007). *Indian J. Agric. Sci.*, 77: 106-108.

Chu, L.C. (1979). In: *Nitrogen and Rice*, IRRI, Philippines, p. 375-418.

Dahatonde, B.B. and Shava, S.V. (1994). *Indian J. Agron.*, 29: 678-680.

Dhaliwal, B.S., Virk, M.S. and Brar, B.S. (2007). *Indian J. Agric. Sci.*, 77: 79-83.

Dhawan, A.S., Zambre, H.A. and Malewar, G.U. (1996). In: *Proc. Natl. Symp. Use of Rock Phosphate for Sust. Agric.* (R. Siddaranappa, M.S. Bedrinath and M.A. Grish eds) USA Bangalore, p. 133-135.

Dobreiner, J. and Day, J.M. (1976). *Proc. Ist. Int. Symp. Nitrogen Fixation* (W.E. Newton and C.J. Nyman eds). Wastigton State Univ. Pullman, p. 518-538.

Dudal, R. and Roy, R.N. (1995). Integrated Plant Nutrient Systems. *FAO Fertilizer and Plant Nution Bull.*, 12: 181-198.

Dwivedi, B.S. and Dwivedi, V. (2007). *Indian J. Ferti.*, 3(1): 11-23.

DWR (2006). Annual Report 2005-06, Directorate of wheat Research, Karnal, India, p. 36.

FAI, Fertilizer Statistics 2006-07, The Fertilizer Association of India, New Delhi.

Ganeshamurthy, A.N., Srinivasarao, Ch., Singh, K.K. and Ali, M. (2003). *Ferti. News*, 48(5): 23-41.

Gaur, A.C. (1989). Proc. Symp. Soil Biol., Hisar (M.M.Mishra and K.K. Kapur eds.), p. 125-128.

Gaur, A.C. (1990). *Phosphate Solubilizing Microorganisms as Biofertilizers*, Omega Sci. Pub., New Delhi.

Ghosh, P.K., Bandyopadgyay, K.K., Manne, H.C., Wanjari, R.H., Misra, A.K. and Subba Rao, A., (2007). *J. Sust. Agric.*, 30: 59-86.

Gill, M.S., Gangwar, B. and Gangwar, K.S. (2008). *Indian J. Fert.*, 4(8): 31-56.

Gupta, J.P. and Rao, A.V. (1989). *Indian J. Agric. Sci.*, 59: 181-182.

Gupta, R.K., Sharma, V.R. and Sharma, K.N. (2002). *Prograssive Fmg. (PAU)*, 38: 22-24.

Gupta, R.K., Yadvinder Singh, Singh, V. and Singh, Bijai (2006). *Indian Fmg.*, 56(4): 10-14.

Gupta, S.B., Katre, R.K., Chhonkar, P.K., Gupta, M.K. and Adil, M.L. (2000). *Fert. News*, 45(1): 35-40.

Heard, J., Cavers, C. and Adrian, G. (2006). *Better Crops*, 90(3): 10-11.

Habbara, M. and Suseeladevi, L. (1990). *Curr. Res.*, 19: 56-57.

Hegde, D.M. (1998). *Indian J. Agric. Sci.*, 68: 144-148.

Hegde, D.M.(1998). *Indian J. Agron.*, 43: 7-12.

Hegde, D.M. (1998). *J. Oilseed Res.*, 15(1): 1-17.

IARI (1994). Status Report 1983-89, AICRP on Renewable Energy Sources for Agric. Agrobased Industry, IARI, New Delhi.

ICARI (1987). Biological Nutrogen Fixation in groundnut- Technologies for Better crops, ICAR, New Delhi; No. 37.

Juwarkar, A.S., Shende, A., Thawale, P.R., Satyanarayanan, S., Deshbratar, P.B. and Bal, A.S. (1994). In: *Fertilizers, Organic Manures, Recycable Wastes and Biofertilizers* (H.L.S. Tandon ed.). FDCO, New Delhi, p. 74.

Jain, P.C., Kushwaha, P.S., Dhakad U.S., Khan, H. and Trivedi, S.K. (1999). *Legume Res.*, 22: 241-244.

Jisha, M.S. and Algawadi, A.R. (1996). *Microbiological Res.*, 151: 213-217.

Joshi, A.B. (2002). *Asian Agri.-History*, 6(1): 43, 62.

Kavimani, R., Annadurai, K. and Vijayabhaskaran, S. *Res. Crops* 1: 348-350.

Kuligod, V., Dasog, G.S. and vijaysekhar, K., IndianJ. *Agric, Sci.* 78, 498-504 (2008)

Kumar, A., Mahapatra, B.S., Mishra, A. Patro, H.K. and Singh, S.P., *Indian J.Agric. Sci.* 77, 273-275 (2005).

Kumar, K., Goh, K.M.Scoyy, W.R. and Farmpton, C.R., *J.Agric. Sci.* 136, 35-53(2001).

Kumar, V. and Ahlawat, IPS., *Indian J. Agric.Sci.* 76, 465-468 (2006).

Kumar, V., Ghosh, B.C. and Bhat, R., *J. Agric. Sci. Camb.* 132, 149-154 (1999).

Kumar, V., Govil, B.P> and Kaore, S.V.(eds.) *Proc. National Wkshp.* IPNS, March 10-12, 1997, Bhubaneswar, IFFCO, New Delhi (1997).

Kundu, KD.K. and Singh, R., Abstr. *Int. Conf. Rice Based Fmg. Syst.* Cuttak November 17-19 2004, p 41-42 (2004).

Ladha, K.C., Sharma, R.K., Sharma, S.K. and Jain, P.M., *Indian J. Dryland. Agric. Res. and Dev.* 21, 177-184 (2006).

Lefroy, R.D.B., Chaitep, W. and Bliar, G.J. *Aust. J. Agric. Res.* 45, 657-667 (1994).

Manjaiah, K. and Singh, D., *Agric. Ecosyst. Environ.* 86, 156-162 (2001).

Manjais, K.M., Channal, H.T., and Satanarayana, T., In *Proc. Natl. Symp.* (R.Sidaramappa, M.S. Badrinath and M.A. Girisheds) UAS Bangalore, p 143-147(1996).

Mankotia, B.S., and Shekhar, J., *Indian Agric. Sci.*, 77.604-606 (2007)

Mankotia, B.S., *Indian J. Agric. Sci.* 77, 604-606 (2007)

Mann, K.K., Brar, B.S. and Dhillon, N.S., *Indian J. Agric. Sci.* 76, 477, 480(2006)

Mathew, J.and Thampatti.K.C.M., *Indian J. Fert.* 3(4), 53-56 (2007).

Mathur, B.S., Sarkar, A.K. and Mishra, B., *J. Indian Soc. Soil Sci.* 28, 206-212 (1980).

Meelu, O.P., Yadvinder Singh, Bijai Singh, Khera, T.S. and Kumar, K., In: *Temperate Rice: Achievents and Potential*, NSW Agriculture, Grifftth, NSW, Australia, Vol. 2 P. 605-613 (1994).

Meelu, O.P., Yadvinder Singh, Bijai Singh, Green Mauring for Soil Productivity Improvement, FAO World Soil Resource Rept. 76, FAO, Rome (1994).

Mehla, D.S., Singh, J.P., Kuhad, I.P., Panwar, D.V.S. and Jalali, B.L. *Nutrient Management Strategies for Rice Production in Haryana*, Dept. of Soil Science and Regional Res. Sat. Karnal, CCS Haryana, Agric. University, Hisar Res. Bull. pp. 52 (2001).

Mishra, B.N., Prasad, R and Gangih, B., *Acta Agron. Hung.* 52, 391-397 (2004).

Mishra, M.M., Kapoor, K.K. and Yadav, K.S., *Indian J. Agric. Sci.* 52, 674-678 (1982).

Mishra, M.M., Khurana, A.L., Dudeja, S.S. and Kapoor, K.K., *Trop. Agric.* 61, 174-178 (1984).

Mishra, V.K. and Saha, R., *Indian J. Agric. Sci.* 78, 399-404 (2004).

Mishra, B.N., Prasad, R and Gangih, B., *Acta Agron. Hung.* 52, 291- 297 (2004).

Mishra, B.N., Prasad, R and Gangih, B. and Siva kumar, B.C., *J. Sust. Agric.* 28, 55- 66 (2006).

Mondal, S.S., Jana, P. and Chandra, P. *Indian J. Agric. Sci.* 78, 534-53 (2008).

Mondal, S.P.L., Gupta, D.A. and Chavan, A.S. *J.Mah. Agric. Univ.* 14, 178-181 (1989).

Nalatwadamath, S.K., Patil, S.K., Adhikari, R.N. and Man Mohan, S. *Indian J. Dryland Agric. Res. and Dev.* 21, 99-104 (2006).

Namdeo, S.L. and Gupta, S.C., In *National Sem. Organic Farming* (M.M. Rai. And L.N. Verma eds). J.N. Krishi Vishwa Vidhalaya, Jabalpur, p. 150-161 (1992).

Narula, N., Kumar, V. and Behl, R.K., *Acta Agron. Hung.* 49, 141-149 (2001).

NATP, Progress Reot. Mid – Term Review – Irrigated Agro-eco systems, Directorate of Maize Research, IARI, New Delhi. P 29.

Nene, Y.L., *Asian Agric.-History* 10, 179-202 (2006).

PAU., AICRP on Renewable Energy Sources of Agriculture and Agrobased Industry, Dept. of Civil Engineering, Punjab Agriculture University, Ludhiana, (1996).

Pal, S.S., Katyal, V. and Gangwar, B., *Indian J. Fert.* 4(7), 65-68 (2008).

Pandey, A. and Kumar, S., *J. Sci. Industr. Res.* 48, 134-144 (1989).

Patil, P.L., *Nat. Sem. Biofertilizers*, Vigyan Bhawan, Oct. 09-10 (1985).

Prabhakaran, J. and Sivasubramanian, K., *Indian J. Agron.*36, 617-618 (1991).

Prasad, M. and Prasad, R., *Indian J. Agron.* 74, 273-275 (2004).

Prasad, R. and Misra, B.N., *Arch. Agron. Soil Sci.* 46, 455-63 (2001).

Prasad, R. and Power, J.F., *Adv. Soil Sci.* 15 209-239 (1991).

Prasad, R. Abstr. *Nat. Symp. Res. Conserv. Agri. Prod.* November 22-25, Ludhiana, p. 299-300(2004).

Prasad, R., *Adv. Agron.* 86, 255-339 (2005).

Prasad, R. *Crop Nutrition*, New Vishal Pub., New Delhi (2007).

Prasad, R., *Curr. Sci.* 89, 252-254 (2005).

Prasad, R., Gangiah, B. and Aipe, K.C. *Exp. Agric.*, 35, 427-435 (1999).

Prasad, R., Sharma, S.N and Singh, S., Summer mung for sustaining rice-wheat Cropping system, Division of Agronomy, IARI, New Delhi, Bulletin (1999).

Prasad, R., Souvenior, 2nd *Int. Agron. Congr.*, November 26-30, 2002, New Delhi (2002).

Prasad, R., Singh, D.K. and R.K., *Arch. Agron. Soil Sci.* 47, 133-139 (2001).

Rachewad, S.N., Raut, R.S. Malewar, G.U. and Ganure, C.K., *J. Mah. Agric. Univ.* 17, 480-481(1992).

Rajput, A.L., *Indian J. Agron.* 40 292-294 (1995).

Ramesh, P. Panwar, N.R., Singh, A.B. and Ramana, S., *Indian J. Agric. Sci.*, 78, 351-354 (2008).

Raverkar, K.P., Gupta, S.B. and Rao, D.L.L., J. *Indian Soc. Soil Sci.*, 53, 591-597 (2005).

Ravi Kumar, B.N., Krishnappa, A.M. and Krishnappa, M., In *Proc. Natl. Symp.Use. of Rock Phosphate for Sust. Agric.* (R.Sidaramappa, M.S. Badrinath and M.A. Girish eds) UAS Bangalore, p 121-125 (1996).

Raddy, B.N. and Sudhakarababu, S.N. *Indian J. Agric. Sci.* 66, 16-19 (1996).

Raddy, N.R.N. and Ahlawat, I.P.S., *Indian J. Agron.* 43, 93-101 (1998).

Roger, P.A. and Kulassoriya, S.A., Blue- green Algae in rice, IRRI, Phillippines. Rout, K.K., Preliminary Rept. Effect of Phosphogypsum on crop productivity, crop quality and environment, OUAT, Bhubaneshwae (2006-07).

Rupela, O.P. and Kumar Rao, J.D.V.K., In *Adoption of Chikpea and Pigeonpea to Abiotic Stresses*, Proc. Consultants Workshop, Dec. 19-21, 1987, ICRISAT, Patancheru, A.P., p 123-131 (1987).

Rupela, O.P., Toomason, B., Mittal, S., Dart, P.J. and Thomson, J.A., *Soil Biol. Biochem.* 19, 247-252 (1987).

Sahu, S.K. and Nayak, S.C., *Ann. Prog. Rept. AICRP on Micro and Secondary Nutrients and Pollutant.*

Sharma, J.S., In *Improving the Productivity and Sustainability of rice-wheat systems: Issues and Impacts*, Am. Soc. Agron. Pub. 65, p. 173-196 (2003).

Sarkar, A., Yadav, R.L., Gangwar, B. and Bhatia, P.C., *Crop Residues in India*, PDCSR, Modipuram (1999).

Sarkar, A.K. and Singh, R.P., *Fert. News* 47 (11), 107-111 (2002).

Sharma, A.R. and Mittra, B.N., *J.Agric. Sci. Camb.* 60, 70-72 (19992).

Sharma, A.R. and Mittra, B.N., *J.Agric. Sci. Camb.* 110, 605-608 (1988).

Sharma, B.K., In *Environmental Chemistry*, Goel Pub. House, Meerut, P. 267-311 (2000).

Sharma, H.L., Singh, C.M. and Modgal, S.C., *Indian J. Agric. Sci.* 57, 163-168 (1987).

Sharma, P.K., Verma T, S. and Bhagat, R.M., *Soil Use Manage.* 11. 199-203 (1995).

Sharma, S.N. and Prasad, R., *Acta Agron. Hung.* 49, 369-378(2001).

Sharma, S.N. and Prasad, R., *Arch. Agron. Soil Sci.* 48, 363-370 (2002).

Sharma, S.N. and Prasad, R., *Bioresource Tech.* 67, 171-175 (1999).

Sharma, S.N. and Prasad, R., *J Agric. Sci.* 141, 359- 369 (2003).

Sharma, S.N., and Prasad, R., *J. Plant Nutr. Soil Sci.* 171, 01-08 (2008).

Sharma, S.N., *Fert. News.* 50 (2), 53- 64 (2005).

Sharma, S.N., *Indian J. Agric. Sci.* 71, 293-295 (2001).

Sharma, S.N., *Indian J. Agric. Sci.* 72, 449 -452 (2002).

Sharma, S.N. and Prasad, R. and Singh, S., *Fert. Res.* 44, 163-168 (1996).

Sharma, S.N. and Prasad, R. and Singh, S., *Plant and Soil* 172, 123-129 (1995).

Sharma, S.N. and Prasad, R. Singh, S. and Singh, P., *J. Agric. Camb.* 134, 169- 172 (2000).

Sindhu, B.S., Beri, V. and Goel, S.R. *Proc. Nat. Symp. Dev. Soil*, Ludhiana November, 2005, (1995), Indian Soc. Soil., New Delhi. P. 44-45 (1995).

Singh, B and Pareek, R.G., *Indian J. Pulses Res.* 16 (1), 31- 33 (2003).

Singh, S, Prasad, R., Singh, B.V., Goel, S.K. and Sharma, S.N. Biol. *Fertility Soils* 9, 235 – 238 (1990).

Singh, A.B. and Tripathi, A.K., *Khad Patrika* (in Hindi) 48 (2), 21-25 (2007).

Singh, K.P., Srivastva, T.K., Singh, P.N. and Archana Suman, *Indian J. Agric. Sci.* 77, 84 -87 (20007).

Singh, P.K., In *Nitrogen and Rice*, IRRI, Philippines P 467- 418 (1979).

Singh, P.K., In *Rice Research in India*, ICAR, New Delhi P. 344-362 (1979).

Singh, R.N., *Role of Blue-Green Algae in Nitrogen Economy of Indian Agriculture*, ICAR, New Delhi (1961).

Singh, S., Singh, R.N.Singh, B.P. and Kushwaha, A.K. *Indian J. Fert.* 4(3), 25-32 (2008).

Singh, S., Prasad, R., Goyal, S.K., Singh, B.V., Varwaha, T.S.and Sharma, S.N., *Indian J.Agron.*, 37, 569-571 (1992)

Singh, S.K., Sharma, M.C., Singh, R. and Singh, M.P., *Indian Fmg.* 56 (9), 16-18

Soni, M.L. and Singh, J.P., *J. Indian Soc. Soil Sci.* 42, 362 – 372 (1994).

Subba Rao, A., Chand, S. and Srivastva, S., *Fert. News* 47(12), 75-90, (2002).

Subba Rao, N.S., In *Cereal Nitrogen Fixation* (S.P. Wani ed.), ICROSAT, Patancheru, A.P. p 23-30(1986).

Surekha, K., Rao, K.V. and Sam, T.K., *Indian J. Agric. Sci.* 78 : 173-176 (2008).

Suri, V.K., Chander, G. Chowdhary, A.K. and Verma, T.S., *Crop Res.* 31, 357 – 361 (2006).

Swarnalakshmi, K., Wattal Dhar, D. and Singh, P.K., *Proc. Indian Nat. Sci. Acad.* 72, 167 – 178.

Swarup, A., *Fert. News* 47 (12), 59 – 73 (2002).

Swarup, A., Reddy, D. and Prasad, R.N. eds, Long-term soil fertility Management through Integrated Plant nutrient Supply, Indian Institute of Soil science, Bhopal. pp. 333 (1998).

Tewatia, R.K., Choudhary, R.S. and Kalwe, S.P., proc. TSI – FAI- IFA Symp. Cum Workshop on Sulphur Balanced Fertilisation, Oct. 4-5, 2006, New Delhi. P 15-25 (2007).

Tewatia, R.K., Kalwe, S.P., and Choudhary, R.K., *Indian J. Fert.* 3 (1) : 111 – 118 (2007)

Tiwari, K.N. and Gupta, B.R, *Indian J. Fert.* 1 (11), 37-52 (2006).

Tiwari, K.N., *Indian J. Fert.* 3(1), 33 – 52 (2007).

Tiwari, K.N., Pathak, A.N. and Lehri, L.K., *Indian J. Agric. Res.* 27, 137 -145 (1993).

Tripathi, S., Joshi, H.C., Sharma, D.K. and Singh, J.P., *Indian J. Agric. Sci.* 77, 208 -211 (2007).

Venkataraman, G.S., *Blue-Green Algae for Rice Production*, FAO, Rome, Bull, 46(1).

Venkateswarlu, B., *Indian J. Microbial*. 44, 215 – 217 (2004).

Venkateswarlu, B., In: *Agriculture Nitrogen use and its Enveronmental Implications* (Y.P. Abrol, N, Raghuraman and M.S. Sachdev eds.) I.K. Int. Pub. House Pvt. Ltd., new Delhi, P 215 -435 (2007).

Venkateswarlu, B., *Indian J. Dryland Agric Res. Dev*. 7, 124 -132 (1992).

Venkateswarlu, B. and Rao, A.V., *Plant and Soil* 74, 379 – 386 (1983).

Venkateswarlu, B. and Rao, A.V., *Biol. Fertility Soils* 5, 18- 22 (1987).

Vineela, S., Wani, S.P., Srinivasa Rao, Ch. Padmaja, B. and Vittal, K.P.P., *App. Soil Ecol*. (2008) (Science Direct dol: 10. 1016/J. apsoil. 2008. 2008.04.001).

Wani, S.P. and Lee, K.K., In *Micro-organisms as Biological Inputs for Sustainable Agriculture* (P.K.Thomson eds.), Peekey Tree Crops and Dev. Foundation, Cochin, p. 39- 76 (1995).

Wani, S.P., *Indian J. Microbiol*. 30, 363 – 393 (1990).

Wani, S.P., Rupela, O.P. and Lee, K.K. *Plant and Soil* 174, 24-49 (1995).

Wankhede, S.Z., Dabre, W.H., Lanjewan, B.K., Sontekey, P.Y.and Takzure, S.C., *Indian J. Agron*., 37, 384-394(1992).

WHO, Health guidelines for the use of Wastewater in agriculture and aquaculture. WHO, Geneva (1989).

Yadav, D.S. and Singh, S.B., *Indian J. Agron*. 31, 380 -383 (1986).

Yadav, R.L., Dwivedi, B.S., Prasad, K.Tomar, O.K, Shrupali, N.J.and Pandey, P.S. *Field Crops Res*., 68, 219-246(2000).

Yadav, R.L., Prasad, S.R., Singh, R. and Srivastava, V.K., *Bioresource Tech*., 49, 231-235 (1994).

Yadav, R.L., Singh, T. and Srivastava, A.K., *Biological Wastes*, 20: 133–141 (1987).

Yadvinder Singh and Meelu, O.P., *Prograssive Fmg (PAU)* 31(5), 12- 13 (1995).

Yadvinder Singh, BijaiSingh, Gupta, R.K.Khind, C.S .and Ladha, J.K., *Int. Rice Res. News* 28 (1), 56- 61 (2003).

Yadvinder Singh, Bijai Singh, and Khind, C.S., *Adv. Soil Sci*. 20, 237-309 (1992).

Yadvinder Singh, Bijai Singh, Maskina, M.S. and Meelu, O.P., *Biol. Fertility Soils* 6, 183-187(1988).

Chapter 14

Crop Residues a Boom for Effective Soil Nutrient Management

Asisan Minz[1] and Rema Das[2]*

[1]Department of Soil Science and Agricultural Chemistry, [2]Department of Agricultural Extension, Birsa Agricultural University, Ranchi – 834 006, Jharkhand

Soil is a very important factor of the plant growth and crop yield. But now a days, very small area of the soil can actually be fertile for agriculture, and if we manage improperly it can be depleted. So the big problem, how we manage and increase the fertility of soil. It has been reported that soil organic carbon and soil organic matter is the most important indicator of soil quality and soil health. It is also beneficial for agricultural sustainability. In this paper, we summarized how crop residue management affects soil chemical parameters. Proper use of crop residue can increase or maintain the physical and chemical properties of SOM and improve the quality of soil. Although, crop residue alone may not be adequate to maintain SOC levels as well as fertility levels of soil. There are several positive effects of retaining crop residues on soil quality, soil organic matter and carbon storage, soil moisture retention, enhanced nutrient cycling, and decreased soil loss, among other environmental and soil health benefits. Negative effects of crop residue retention attributed to nitrogen immobilization, water logging, decreased soil temperature and attack of termites. The incorporation of crop residue can save and substitute inorganic potassic fertilizer up to some extent and it also improve soil health.

Keywords: *Crop residues, Soil nutrients, Soil organic carbon, Soil pH.*

Food security, nutritional security, sustainability and profitability are the main foci of present day agriculture in India. In the absence of any possibility of bringing additional land under plough, vertical growth in agriculture through intensive and multiple cropping is the only alternative. Intensification of agriculture to meet the growing needs of the growing population has brought with it several soil and crop related constraints. A widespread nutrient deficiency in soil has raised doubts on its capacity to sustain additional production.

The Green Revolution in India which took place in the 1960's was responsible for increasing agricultural productivity and overall food production in the country through application of high yielding verities (HYVs), high analysis chemical fertilizers and chemical insecticides and pesticides along with huge application of ground water in indiscriminate ways. As a result, India had a surplus stock of grain production in cereals. No doubt chemical fertilizers provide nutrients in readily available form. But its continuous use in a wide variety of soils of India, reveal that it deteriorates the soil health and leads to deleterious effect on a long term basis if used indiscriminately and in an imbalanced proportion. Apart from the soil productivity, the use of chemical fertilizers is also becoming costly due to its scarcity during peak demand period of crops. There are indications of stagnation or even decline in the productivity of this cropping system due to decline in soil organic matter, over mining of nutrient reserve, loss of nutrients and non-availability of cost effective fertilizers. Further, the application of inorganic fertilizers even in balanced form may not sustain soil fertility under continuous cropping. Soil fertility experiments have shown the need of blending the use of inorganic fertilizers with crop residues or organic sources for better soil health in the long run.

However, integrated use of inorganic and organics including crop residues may improve the soil productivity in the long run. In the coming decades, a major issue in designing sustainable agriculture system will be the management and rational use of organic inputs such as green manure and crop residues for crop production (Powlson, 1994). Crop residues provide a number of environmental services when left in the field, including contributing to the formation of SOC, preventing erosion, reducing evaporation from the soil surface, improving soil structure, supporting living organisms, contributing nutrients to the soil, and providing water filtration and retention capacity (Powlson *et al.*, 2011; Lal, 2014; *Nicholson et al.*, 2014; Johnston *et al.*, 2009).

Crop residue incorporation is one way to maintain the fertility of the soil on the farm. Crop residue is not a waste but rather a good natural resource. About 25 per cent of nitrogen (N) and phosphorus (P), 50 per cent of sulfur (S), and 75 per cent of potassium (K) uptake by cereal crops are retained in crop residues, making them valuable nutrient sources for crops (Kumar *et al.*, 2015, Singh *et al.*, 2018). Crop residues retain considerable quantity of plant nutrients and can be utilized as good source of plant nutrients, primary source of organic matter and important components for the stability of agricultural ecosystems (Kabirinejad *et al.*, 2014). Recycling of crop residues is one of the ways of improving soil nutrient content and maintaining soil productivity (Beres and Kazinczi, 2000). Due to the limited availability of alternative organic amendments, now a days, farmers are considering the incorporation of crop residue in fields can be considered key in promoting physical, chemical, and biological soil health (Marie-Soleil Turmel *et al.*, 2014). However, due to multiple purpose uses, small landholders are faced with trade-offs in managing crop residues in farm. The incorporation of crop residue include increase in carbon storage, soil quality, soil organic matter, enhanced nutrient release rate, minimize soil erosion, reduced weed control and other environmental health benefits. The rate of nutrient release not only depends on the technique followed, but also on the quality of crop residue. Integrated soil fertility management is a new approach to raising productivity levels while maintaining natural resources (crop residues/stables/straw). Integrated soil fertility management focuses on restoring soil nutrient pools, reducing nutrient losses, and improving efficiency. The cycle of sustainability must facilitate the return of all farm resources back to the field.

Soil Nutrient Management

The soil nutrient management is the practice of managing the application of commercially available fertilizers, manures, soil amendments, and other organic by-products to agricultural farm land as a source of plant nutrients. Crop straw incorporation improves soil organic carbon and soil nutrients contents. It is beneficial for recycling nutrients residue, ploughing is important in immobilization of nutrients (especially nitrogen), and the better ratio of C:N needs to be corrected by applying extra nitrogen fertilizer at the time of residue in-corporation (Yadvinder Singh *et al.*, 2005).

1. Soil Organic Carbon

Soil organic carbon (SOC) is considered as an important indicator of soil quality and agricultural sustainability because it improves soil aggregate stability and soil water retention, and provides a reservoir of soil nutrients (Liu *et al.*, 2006). Soil organic carbon increasing soil organic matter content in soil is an important index to measure for improved soil health quality. Now a days due to mismanagement agricultural practices disturb the SOC pool, which represents a large potential to loss soil C thus lead to lower availability of soil nutrients, and ultimately reduces soil quality. The straw incorporation significantly increased the organic carbon concentration and storage levels as compared to without incorporation of straw, where the increase was probably associated with the amount of crop residues incorporated into the soil. The soil nutrient availability directly depend on soil organic matter content, the soil organic matter can be divided into the labile pool and the humus pool. The labile pool means those organic resources which are easily decomposed by soil microorganisms, while the humus pool is more resistant to decomposition. The decomposition of locally available natural farm resources can transform into soil organic matter through naturally occurring soil microbes, that can biochemically transform raw residue materials into humus, this humified materials act as a stores house of different nutrients and releases energy and nutrients that, in turn, become available for plant uptake (Minz and Das 2020). Rakshit *et al.* (2016b) found that the application of recommended 100 per cent NPK to super-optimal doses (200 per cent NPK) of mineral fertilizers on root growth pattern, microbial biomass and counts, organic carbon and mineral nitrogen at various phenological stages of wheat growth. Soil organic carbon (SOC) observed at 200 per cent NPK (0.375 per cent) was higher than at 100 per cent NPK (0.335 per cent), but it was non-significant. SOC content in soil in the treatments at anthesis stage was significantly higher than those in the treatments at CRI and maturity stage. Amongst the various stages of wheat growth, anthesis was most conspicuous, where the values of SOC were found to be higher (mean of 0.392 per cent over the two cropping cycles). The value of SOC varied from 0.349 per cent at CRI and 0.336 per cent at maturity over the two cropping cycles. It was found to be higher at anthesis stage (16.67 per cent and 12.32 per cent higher over maturity and Crown Root Initiation (CRI) stage, respectively). Bhavya *et al.* (2018) revealed that the organic carbon content of soil was higher in the surface layer of the soil *i.e.*, 0-15 cm as compared to other subsurface soil depth in all the cropping systems. This is due to the continuous addition of organic matter which leads to accumulate more organic matter to the soil and thereby it helps to increase the carbon content in soil. Since crop residues are incorporated in to the surface soil and a major portion of the left over residues of shallow rooted crops usually accumulate in the top few centimetre of the soil layers, there was possibility for a relatively greater accumulation of organic carbon in 0-15 cm soil as compared to the soils of lower layer.

Crop residue retention is key to increasing and/or maintaining SOC levels; however, its effect may be controlled by soil type, climate and management factors (Govaerts *et al.*, 2009b). In addition to nutrients, soil organic matter can improve soil texture, structure, and chemical balance, while providing habitat for a greater diversity of soil flora and fauna. Soil management techniques that are focused on increasing soil organic matter will tend to lead to systems with adequate nutrient supplies and storage capacity. These facts clearly indicates that without hampering soil we can get more food and ultimately can also achieved food security through sustainable agriculture.

2. Soil pH

The response of crop residue incorporation on soil pH is directly dependent on the quality of the residue incorporation and types of soils. The incorporation of crop residues can affect soil pH due to direction of change in chemical composition of the residue and soil properties (Butterly *et al.*, 2011). Changes in pH from crop residue addition are correlated with the release of total concentration of organic anions and the nitrogen content in residue (Tang and Yu, 1999; Butterly *et al.*, 2011) after microbial decomposition. Yadav and Gupta (2018) conducted an experiment and find out on an average, pH ranged from 7.64 to 8.45 with mean value of 8.1 in surface (0-15cm) soils. However, it varies from 7.87 to 8.65 with mean value of 8.3 in upper subsurface (15-30 cm) and from 7.86 to 8.78 with mean value of 8.4 in lower subsurface (30-60 cm) soils. The higher pH in lower layers could be due to increase in accumulation of exchangeable of cations. The increase in soil pH with increase in soil depth was also reported by Yadav *et al.* (2016).

Bhatt *et al.* (2019) results revealed that the application of NPK along with organic residues increased the pH and organic carbon. In all the treatments of sub-subsurface layer nutrient status decreased as compared to surface layer except in case of pH. The soil pH and organic carbon ranged from 7.24 to 7.64 and 0.58 to 1.12 per cent, respectively at the surface layer (0-15 cm) and 7.58 to 7.77 and 0.27 to 0.61 per cent, respectively at the sub-surface layer (15-30 cm). The soil pH increased with increase in soil depths. This might be due to leaching of soluble salt from surface to sub-surface soil.

The decomposition of organic anions contributes to the change in soil pH by two mechanisms: (1) decarboxylation of organic anions that consume H^+ anions; if the initial soil pH is less than the pKa of the weak acid group of the organic carbon then the H^+ will associate with the organic anions and vice versa (Bonn and Fish, 1991), and (2) ligand exchange with hydroxyl groups of aluminum and iron oxides. Mineralization of nitrogen contributes to changes in soil pH by initially consuming H^+ during the ammonification process and subsequently releasing H^+ during nitrification. Due to disequilibrium in the nitrogen cycle the overall effect is acidifying. Thus the final soil pH is determined by the balance of these reactions (Xu and Coventry, 2003).

3. Available Nitrogen

Addition of crop residues directly influences the availability of soil nitrogen to the crop. The incorporation of leguminous crop residues can result nitrogen mineralization this might be due to having a low C/N ratio, whereas cereal residues with a high C/N composition can leads temporarily immobilize N during the decomposition process (Govaerts *et al.*, 2006b; Aulakh *et al.*, 1991) at initial stages of crop growth. Azmi *et al.* (2019) reveals that Available N and organic carbon content decreased regularly with increasing soil depth, irrespective of treatments. Maximum available N was recorded in the treatment receiving inorganic fertilizers + compost + crop residues and minimum available N was recorded in control plot at all the soil depths. The average available N (301kg ha^{-1}) was recorded maximum in the surface layer (0-15 cm) whereas; minimum available N (99 kg ha^{-1}) was noted in the 90-120 cm soil depth. The maximum organic carbon content was recorded in surface soil (4.0 to 9.8 g kg^{-1}) under all treatment. Correlation study showed that available N and organic carbon content in soil were positively and significantly correlated with grain yield and N uptake by rice and wheat. An experiment was conducted during rabi, 2008-09 at Research Farm of Allahabad Agricultural Institute, Uttar Pradesh in an alluvial soil. The highest available N (206.7 kg ha^{-1}), P (14.6 kg ha^{-1}) and K (212.7 kg ha^{-1}) values were recorded with addition of straw in combination with green manure while, the lowest values were associated with inorganic fertilizer treated plots (170.4, 12.4 and 207.2 kg ha^{-1} of N, P and K, respectively) in wheat crop (Dhar *et al.*, 2014).

The loss of soil mineral nitrogen is also higher by denitrification processes of applied fertilizer can also be greater when crop residues are directly left on the surface, it might be due to higher soil moisture (Aulakh *et al.*, 1984). Negative effects of crop residue retention are less common; however, they have been reported in some environments. The most common negative effect is N mobilization which can reduce N availability to the crop in N limiting environments.

4. Available Phosphorus

Residue retention has been found to increase the concentration of P in the top soil. This can be attributed to redistribution of P mined from the lower soil layers (Zibilske *et al.*, 2002). Jat *et al.* (2017) a field experiment was conducted for 5 year to study the effect of crop residue management practices alone or with combination of FYM on maize (*Zea mays*) crop. Results revealed that the incorporation of crop residues and/or FYM increased available P content of soil over the control after five years of study. Aggarwal *et al.*, 1997, ascribed the increase in available P under residues incorporation due to increase in phosphatase (both acid and alkaline) enzyme activity. The incorporation of crop residues can indirectly increase the availability of soil phosphorus, it might be due to several humic molecules and low molecular weight aliphatic acids are released during the decomposition of incorporated crop residues, which can block Al-oxide adsorption sites and reduce overall adsorption of P (Haynes and Mokolobate, 2001). Gupta *et al.* (2007) reported that P sorption was significantly reduced with the incorporation of crop residue compared with removal of residues. Phosphorus release was 21-32 per cent higher with straw incorporation or residue burned than when it was removed.

5. Available Potash

The incorporation of crop residues supplied the soil with significant amounts of potassium. Crop residues contain large quantities of potassium, and their recycling can markedly increase potassium availability in soils (Mubarak and Dawi 2009). Bai *et al.* (2015) reported that straw return significantly increased K absorption by the crops, with significant increase in straw part. In treatment with straw return, the K content in crop straw increased by 15.9 and 21.8 per cent in wheat and maize, respectively, compared with no straw return treatment. But, straw return had also a little effect on K content in grain of the crops. Straw return had significant influences on total K uptake by wheat and maize plants, with an increase of 32.7 and 30.9 per cent, respectively, compared with no straw return treatment. The contents of soil available K and soil organic matter were significantly affected by the straw return with an increase of 6.07 and 23.0 per cent, respectively, compared to no straw return treatment. Permanent raised beds with residue retention had 1.65 and 1.43 times higher K concentration in 0-5 cm and 5-20 cm surface layer, respectively, than conventional tilled raised beds with residue retention. The differences in K concentration were decreasing with increasing in depth (Govaerts *et al.*, 2007). The incorporation of crop residue can save and substitute inorganic potassic fertilizer upto some extent and it also improve soil health.

Conclusions

It is inferred that incorporation of crop residue in soil is not only increasing the yield, but it also enhances the other soil fertility parameters, which is necessary to achieve a sustainable production and minimize the depletion of nutrient content of the soil for the longer period. However, the available nutrients in crop residue incorporated plots were slightly higher than without crop residue incorporation. The incorporation of crop residue can save and substitute inorganic potassic fertilizer up to some extent and it also improve soil health.

REFERENCES

Aggarwal, R.K.; Kumar, P. and Power, J.F. (1997). Use of crop residue and manure to conserve and enhance nutrient availability and pearl millet yields in an arid tropical region. *Soil and Tillage Research*, 41: 43- 51.

Aulakh, M.A., Doran, J.W., Walters, D.T., Mosier, A.R. and Francis, D.D. (1991). Crop residue type and placement effects on denitrification and mineralization. *Soil Sci. Soc. Am. J.* 55, 1020–1025.

Aulakh, M.S., Rennie, D.A. and Paul, E.A. (1984). Gaseous nitrogen losses from soils under zero-till as compared with conventional till management systems. *J. Environ. Qual.* 113, 130–136.

Azmi, N.Y., Kumar, M., Seema and Kumari, S. (2019). Long-term effect of fertilizers and organics on depth-wise distribution of available nitrogen and organic carbon under rice-wheat sequence in calcareous soil of Bihar, Global Perspective in Agricultural and Applied Sciences for Food and Environmental Security, International Conference, Uttrakhand,1-2 Dec. 2019, ISBN: 978-93-5391-939-9, 2: 287.

Bai, Y., Wang, L., LU, Y., Yang, L., Zhou, L., Cheng, N.L. and Ming, F. (2015). Effects of long-term full straw return on yield and potassium response in wheat-maize rotation. *Journal of Integrative Agriculture*, 14(12): 2467- 2476.

Beres, I. and Kazinczi, G. (2000). Allelopathic Effects of Shoot Extracts and Residues of Weeds on Field Crops. *Allelopathy Journal*, 7: 93-98.

Bhatt, M., Singh, A.P., Singh, V., Kala, D.C. and Kumar, V. (2019). Long-term effect of organic and inorganic Fertilizers on soil physico-chemical properties of a silty clay loam soil under rice-wheat cropping system in Tarai region of Uttarakhand. *Journal of Pharmacognosy and Phytochemistry*, 8(1): 2113- 2118.

Bhavya, V.P., Kumar, A.S., Kiran, S.K., Alur, A., Shivakumar, K.M. and Shivanna, M. (2018). Effect of Different Cropping System on Important Soil Enzyme Activity, Organic Carbon and Microbial Activity with Different Depth. *Int. J. Curr. Microbiol. App. Sci.*, 7(1): 315-322.

Bonn, B.A. and Fish, W. (1991). Variability in the measurement of humic carboxyl content. *Environ. Sci. Technol.* 25, 232–240.

Butterly, C.R., Kaudal, B.B., Baldock, J.A. and Tang, C. (2011). Contribution of soluble and insoluble fractions of agricultural residues to short-term pH changes. *Eur. J. Soil Sci.* 62, 718–727.

Dhar, D., Ashim, D., Nirmalendu, B., Niladri, P., Shrikant, B. and Thomas, T. (2014). Residual effect of crop residues on growth, yield attributes and soil properties of wheat under rice-wheat cropping system. *Indian Journal of Agriculture Research.* 48(5): 373-378.

Govaerts, B., Sayre, K.D., Ceballos-Ramirez, J.M., Luna-Guido, M.L., Limon-Ortega, A., Deckers, L., Dendooven, L. (2006b). Conventionally tilled and permanent raised beds with different crop residue management: effects on soil C and N dynamics. *Plant Soil* 280, 143–155.

Govaerts, B., Verhulst, N., Castellanos-Navarrete, A., Sayre, K.D., Dixon, J., Dendooven, L. (2009b). Conservation agriculture and soil carbon sequestration: between myth and farmer reality. *Crit. Rev. Plant Sci.*, 28: 97–122.

Govaerts, B., Sayre, K.D., Lichter, K., Dendooven, L. and Deckers, J. (2007). Influence of permanent raised bed planting and residue management on physical and chemical soil quality in rain fed maize/wheat systems. *Plant Soil*, 291: 39-54.

Gupta, R.K., Singh, Y., Ladha, J.K., Singh, B., Singh, J., Singh, G. and Pathak, H. (2007). Yield and phosphorus transformations in a rice-wheat system with crop residue and phosphorus management. *J American Soc Soil Sci.* 71: 1500-1507.

Haynes, R.J., Mokolobate, M.S. (2001). Amelioration of Al toxicity and P deficiency in acid soils by additions of organic residues: a critical review of the phenomenon and the mechanisms involved. *Nutr. Cycl. Agroecosyst*. 59, 47–63.

Jat, M. L., Balyan, J. K., Kumar, V., Rathi, Y. S., Chary, G.R. and Sharma, R.K. (2017). Crop Residue Management for Sustainable Production of Maize (*Zea mays*) in Dryland Ecosystem. *Chem Sci Rev Lett*, 6(23), 1681-1686.

Johnston, E., Poulton, P. R., and Coleman, K. (2009). Soil organic matter: Its importance in sustainable agriculture and carbon dioxide fluxes. In *Advances in Agronomy, Vol. 101*.

Kabirinejad, S, Kalbasi, M., Hossein, A., Khoshgoftar, M., Hoodaji, M. and Afyuni, M. (2014). Effect of Incorporation of Crops Residue into Soil on Some Chemical Properties of Soil and Bioavailability of Copper in Soil. *Int. J. Adv. Biol. Biom. Res,*, 2 (11), 2819-2824.

Kumar, P., Kumar, S. and Joshi, L. (2015). The extent and management of crop stubble. In: Kumar, P, Kumar S, Joshi L (eds) Socio–economic and environmental implications of agricultural residue burning. Springer, Berlin, Germany, 13-34.

Lal, R. (2014). Soil carbon management and climate change. In Hartemink, A., and McSweeney, K. (Eds.), *Soil Carbon*. Retrieved from https: //link.springer.com/ chapter/10.1007/978-3- 319-04084-4_35.

Liu, X., Herbert, S.J., Hashemi, A.M., Zhang, X., Ding, G. (2006). Effects of agricultural management on soil organic matter and carbon transformation – a review. *Plant Soil Environ*. 52, 531–543.

Marie-Soleil T, Speratti A, Baudron F, Verhulst N, Govaerts BL, (2014). Crop residue management and soil health: A systems analysis. *Agril. Syst.*, 134 (2015): 6-16.

Minz, A. and Das, R. (2020). Sustainable Agriculture : A Prerequisite for Food Security. Financial Inclusion and Economic Growth. International Publications, pp.- 229-243. ISBN : 978-81-945988-3-1

Mubarak, A.R. and Dawi, B.S. (2009). Decomposition and nutrient release from various tree litters in a sandy soil of a semi-arid tropics. *Sudan Journal Desert Research*, 1: 36–55.

Nicholson, F., Kindred, D., Bhogal, A., Roques, S., Kerley, J., Twining, S., and Ellis, S. (2014). *Straw incorporation review* (Research Review 81). Home-Grown Cereals Authority. Retrieved from https: //cereals.ahdb.org.uk/publications/2014/ july/30/straw- incorporation-review.aspx.

Powlson, D.S., Glendining, M.J., Coleman, K., and Whitmore, A.P. (2011). Implications for soil properties of removing cereal straw: Results from long-term studies. *Agronomy Journal*, 103(1): 279-287.

Powlson, D.S. (1994). Quantification of nutrient cycles using long-term experiments. Pages 97- 115. In: *Long Term Experiments in Agricultural and Ecological Sciences* (Leigh, R.A. and Johnson, A.E. eds.) Wallingford, U.K., CAB International.

Rakshit, R., Patra, A.K., Purakayastha, T.J., Singh, R.D., Pathak, H., Dhar, S. and Das, A. (2016b). Super-optimal fertilization affects root growth and soil microbial abundance and biomass during wheat growth. *Journal of Food, Agriculture and Environment*. 14(1): 112-118.

Singh, V. K., Dwivedi, B.S., Singh, S.K., Mishra, R.P., Shukla, A.K. and Rathore, S.S. (2018). Effect of tillage and crop establishment, residue management and K fertilization on yield, K use efficiency and apparent K balance under rice-maize system in north-western India. *Field Crops Res.*, 224: 1–12.

Tang, C. and Yu, Q. (1999). Impact of chemical composition of legume residues and initial soil pH on pH change of a soil after residue incorporation. *Plant Soil*, 215: 29–38.

Xu, R.K. and Coventry, D.R. (2003). Soil pH changes associated with lupin and wheat plant materials incorporated in a red-brown earth soil. *Plant Soil*, 250: 113–119.

Yadav, B.K. and Gupta, N. (2018). Enzyme Activities and Nutrient Status in Soil under Ziziphus mauritiana in Semi-Arid Region of Punjab, India. *Int. J. Curr. Microbiol. App. Sci.*, 7(1): 479-490.

Yadav, B.K., Sidhu, A.S. and Thaman, S. (2016). Soil fertility status of Punjab Agricultural University seed farm, Chak Ruldu Singh Wala, Sangat, Bathinda, Punjab. *Annals of Plant and Soil Research*, 18: 226-231.

Yadvinder-Singh, B. S. and Timsina, J. (2005). Crop residue management for nutrient cycling and improv-ing soil productivity in rice-based cropping systems in the tropics. *Advances in Agronomy*, 85: 269-407.

Zibilske, L.M., Bradford, J.M., Smart, J.R. (2002). Conservation tillage induced changes in organic carbon, total nitrogen and available phosphorus in a semi-arid alkaline subtropical soil. *Soil Till. Res.*, 66: 153–163.

Chapter 15

Organic Farming: A Key to Sustainable Agriculture

Meenakhi Prusty[1], Monika Ray[2] and Gyanaranjan Sahoo[3]*

[1]*Regional Research and Technology Transfer Station (RRTTS), Dhenkanal, Odisha*
[2]*Regional Research and Technology Transfer Station (RRTTS), Keonjhar, Odisha*
[3]*Krishi Vigyan Kendra (KVK), Angul, Odisha*

Agriculture is the backbone of Indian economy. The advancement of technology and development in agriculture has enabled our country to provide food security. As every technology has pros and cons; this advancement in agriculture has directed to imbalance our ecosystem by unsystematic application of an enormous quantity of chemical fertilizers, pesticides in terms of their negative impact on the human health and the environment. So Organic Agriculture' is the only solution to nurture the land and to regenerate the soil by going back to our traditional method of farming i.e., free from chemicals, pesticides and fertilizers. Adaptation and mitigation based on organic agriculture can build on the well-established practice because organic agriculture avoids nutrient exploitation and increases soil organic matter content. Consequently, soils under organic farming capture and store more water than soils under conventional cultivation. Furthermore, organic farming reduces the vulnerability of the farmers to climate change and variability by comprising highly diverse farming systems and thus, increases the diversity of income sources and the flexibility to cope with adverse effects of climate change and variability, such as changed rainfall patterns. So this chapter provides a brief outlook about Organic Agriculture, its major components, its present scenario in India, Govt. policies, the main principles of organic agriculture and limitations in practicing organic agriculture.

***Keywords**: Organic agriculture, Sustainable agriculture, Food security, Modern agriculture.*

Agriculture facilitates to meet the indispensable needs of human civilization by providing food, clothing, shelter, medicine and recreation. Hence, agriculture is the most important venture in the world. India's agricultural sector is still very important and considered to be the backbone of Indian economy. India has made significant advances in agricultural production in recent decades, including the introduction of high-yield seed varieties and increased use of fertilizers.

In 1960, Green Revolution (GR) technologies, supported by policies, and fuelled by agrochemicals, machinery and irrigation, are known to have enhanced agricultural production and productivity. While these technologies greatly helped to address the food security of India, farmers using these technologies have to depend upon the purchased inputs. The manufactures of fertilizers and pesticides, the two major inputs of GR technologies, need fossil fuels and/or expensive energy, and are associated with serious environmental and health problems.

Modern agricultural farming practices, along with irrational use of chemical inputs over the past four decades have resulted in not only loss of natural habitat balance and soil health but have also caused many hazards like soil erosion, decreased groundwater level, soil salinization, pollution due to fertilizers and pesticides, genetic erosion, ill effects on environment, reduced food quality and increased the cost of cultivation, rendering the farmer poorer year by year (Ram, 2003). Farmers do not find agriculture a viable proposition anymore and in fact, a large number of farmers have committed suicides (Deshpande, 2002).

Sustainable development has caught the imagination and action all over the world for more than a decade. Sustainable agriculture is necessary to attain the goal of sustainable development. According to the Food and Agriculture Organization (FAO), sustainable agriculture "is the successful management of resources for agriculture to satisfy changing human needs while maintaining or enhancing the quality of environment and conserving natural resources". All definitions of sustainable agriculture lay great emphasis on maintaining an agriculture growth rate, which can meet the demand for food of all living things without draining the basic resources.

Organic farming system in India is not new and is being followed from ancient time. It is a method of farming system which primarily aimed at cultivating the land and raising crops in such a way, as to keep the soil alive and in good health by use of organic wastes (crop, animal and farm wastes, aquatic wastes) and other biological materials along with beneficial microbes (bio fertilizers) to release nutrients to crops for increased sustainable production in an eco-friendly pollution free environment.

Dr. A.P.J. Abdul Kalam stated "Organic agriculture, a holistic system that focuses on improvement of soil health, use of local inputs and relatively high intensity use of local labour, is an admirable fit for dry lands in many ways and the dry land offer many benefits that would make it relatively easy to implement."

FAO suggested that "Organic agriculture is a unique production management system which promotes and enhances agro-ecosystem health, including biodiversity, biological cycles and soil biological activity, and this is accomplished by using on-farm agronomic, biological and mechanical methods in exclusion of all synthetic off-farm input.

The most popularly accepted definition of organic farming is: 'Organic agriculture is a holistic production management system which promotes and enhances agro-ecosystem health, including biodiversity, biological cycles, and soil biological activity (Reddy, 2010). It emphasizes the use of management practices in preference to the use of off-farm inputs, taking into account that regional conditions require locally adapted systems (Reddy, 2010). This is accomplished by using wherever possible, agronomic, biological and mechanical methods, as opposed to using synthetic materials, to fulfil any specific function within the system (FAO, 1999; Reddy, 2010).

What is Organic Agriculture?

Organic agriculture is holistic production management system which promotes and enhances agro-ecosystem health, including biodiversity, biological cycles, and soil biological activity. Organic production systems are based on specific and precise standards of production which aim at achieving optimal agro-ecosystems which are socially, ecologically and economically sustainable.

IFOAM defines organic agriculture as: ¯a production system that sustains the health of soils, ecosystems and people. It relies on ecological processes, biodiversity and cycles adapted to local conditions, rather than the use of inputs with adverse effects. Organic agriculture combines tradition, innovation and science to benefit the shared environment and promote fair relationships and a good quality of life for all involved." The International Federation of Organic Agriculture Movements (IFOAM) has formulated four broad principles of organic farming, which are the basic roots for organic agriculture growth and development in a global context.

1. **Principle of Health***:* Organic agriculture should sustain and enhance the health of soil, plant, animal, human and planet as one and indivisible. Health is the wholeness and integrity of living systems. It is not simply the absence of illness, but the maintenance of physical, mental, social and ecological well-being.
2. **Principle of Ecology***:* Organic agriculture should attain ecological balance through the design of farming systems, establishment of habitats and maintenance of genetic and agricultural diversity. Those who produce, process, trade, or consume organic products should protect and benefit the common environment including landscapes, climate, habitats, biodiversity, air and water.
3. **Principle of Fairness**:Organic agriculture should build on relationships that ensure fairness with regard to the common environment and life opportunities. Fairness is characterized by equity, respect, justice and stewardship of the shared world, both among people and in their relations to other living beings.
4. **Principles of Care***:* Organic agriculture should be managed in a precautionary and responsible manner to protect the health and well-being of current and future generations and the environment. It should prevent significant risks by adopting appropriate technologies and rejecting unpredictable ones, such as genetic engineering.

Main Principle of Organic Farming

- ☆ Protect the environment, minimize soil degradation and erosion, decrease pollution, optimize biological productivity and promote a sound state of health.
- ☆ Maintain long-term soil fertility by optimizing conditions for biological activity within the soil.
- ☆ Maintain biological diversity within the system.
- ☆ Recycle materials and resources to the greatest extent possible within the enterprise.
- ☆ Provide attentive care that promotes the health and meets the behavioural needs of livestock.
- ☆ Prepare organic products, emphasizing careful processing, and handling methods in order to maintain the organic integrity and vital qualities of the products at all stages of production.
- ☆ Rely on renewable resources in locally organized agricultural systems.

Nutrient Management in Organic Farming

In organic farming, it is important to constantly work to build a healthy soil that is rich in organic matter and has all the nutrients that the plants need. Several methods *viz.* green manuring, addition of manures and bio fertilizersetc. can be used to build up soil fertility. These organic sources not only add different nutrients to the soil but also help to prevent weeds and increase soil organic matter to feed soil microorganisms. Soil with high organic matter resists soil erosion, holds water better and thus requires less irrigation. Some natural minerals that are needed by the plants to grow and to improve the soil's consistency can also be added. Soil amendments like lime are added to adjust the soil's pH balance. However soil amendment and water should contain minimum heavy metals. Most of the organic fertilizers used are recycled by-products from other industries that would otherwise go to waste. Farmers also make compost from animal manures and mushroom compost. Before compost can be applied to the fields, it is heated and aged for at least two months, reaching and maintaining an internal temperature of 130°-140°F to kill unwanted bacteria and weed seeds. A number of organic fertilizers/amendments and bacterial and fungal bio fertilizers can be used in organic farming depending upon availability and their suitability to crop. Different available organic inputs are described below:

1. Organic Manures

Commonly available and applied farm yard manure (FYM) and vermicompost *etc.* are generally low in nutrient content, so high application rates are needed to meet crop nutrient requirements. However, in many developing countries including India, the availability of organic manures is not sufficient for crop requirements; partly due to its extensive use of cattle dung in energy production. Green manuring with Sesbania, cowpea, green gram *etc.* are quiet effective to improve the organic matter content of soil. However, use of green manuring has declined in last few decades due to intensive cropping and socioeconomic reasons. Considering these constraints

International Federation of Organic Agriculture Movement (IFOAM) has approved the use of some inorganic sources of plant nutrients like rock phosphate, basic slag, rock potash *etc.* in organic farming systems. These substances can supply essential nutrients and may be from plant, animal, microbial or mineral origin and may undergo physical, enzymatic or microbial processes and their use does not result in unacceptable effects on produce and the environment including soil organisms.

2. Bacterial and Fungal Biofertilizers

Contribution of biological fixation of nitrogen on surface of earth is the highest (67.3 per cent) among all the sources of N fixation. Following bacterial and fungal bio fertilizers can be used as a component of organic farming in different crops.

- ☆ **Rhizobium:**The effectiveness of symbiotic N_2 fixing bacteria *viz.* Rhizobia for legume crops *e.g. Rhizobium, Bradyrhizobium, Sinorhizobium, Azorhizobium,* and *Mesorhizobiumetc*have been well recognized. These bacteria infecting legumes have a global distribution. These rhizobia have a N_2-fixing capability up to 45-60 kg N ha^{-1} depending on host- plant species and bacterial strains. Carrier based inoculants can be coated on seeds for the introduction of bacterial strains into soil.
- ☆ **Azotobacter***:* N_2 fixing free-living bacteria can fix atmospheric nitrogen in cereal crops without any symbiosis. Such free living bacterias are: *Azotobacter*sp. for different cereal crops; *Acetobacterdiazotrophicus*and *Herbaspirillum*spp. for sugarcane, sorghum and maize crop. Besides fixing nitrogen, they also increase germination and vigour in young plants leading to an improved crop stand. They can fix 15-20 kg/ha nitrogen per year. *Azotobacter*sp. also has ability to produce anti-fungal compounds against many plant pathogens. *Azotobacter*can biologically control the nematode diseases of plants also.
- ☆ **Azospirillum***:* The genus *Azospirillum*colonizes in a variety of annual and perennial plants. Studies indicate that *Azospirillum* can increase the growth of crops like sunflower, carrot, oak, sugarbeet, tomato, pepper, cotton, wheat and rice. The crop yield can increase from 5-30 per cent. Inoculum of *Azotobacter* and *Azospirillum* can be produced and applied as in peat formulation through seed coating. The peat formulation can also be directly utilized in field applications.
- ☆ **Plant growth promoting rhizobacteria**: Various bacteria that promote plant growth are collectively called plant growth promoting rhizobacteria (PGPR). PGPR are thought to improve plant growth by colonizing the root system and pre-empting the establishment of suppressing deleterious rhizosphere microorganisms on the roots. Large populations of bacteria established in planting material and roots become a partial sink for nutrients in the rhizosphere thus reducing the amount of C and N available to stimulate spores of fungal pathogens or for subsequent colonization of the root. PGPR belong to several genera *viz.Actinoplanes, Azotobacter, Bacillus, Pseudomonas, Rhizobium, Bradyrhizobium, Streptomyces, Xanthomonas etc. Bacillus* spp. act as biocontrol agent because their

endospores are tolerant to heat and desiccation. Seed treatment with *B.subtilis*is reported to increase yield of carrot by 48 per cent, oats by 33 per cent and groundnut upto 37 per cent.

- **Phosphorus-solubilizing bacteria (PSB):** Phosphorus is the vital nutrient next to nitrogen for plants and microorganisms. This element is necessary for the nodulation by *Rhizobium* and even to nitrogen fixers, *Azolla* and BGA. The phospho microorganism mainly bacteria and fungi make available insoluble phosphorus to the plants. It can increase crop yield up to 200-500 kg/ha and thus 30 to 50 kg Super Phosphate can be saved. Most predominant phosphorus-solubilizing bacteria (PSB) belong to the genera *Bacillus* and *Pseudomonas*. At present PSB is most widely used bio fertilizer in India. PSB can reduce the P requirement of crop up to 25 per cent.
- **Mycorrhizal fungi***:* Root-colonizing mycorrhizal fungi increase tolerance of heavy metal contamination and drought. Mycorrhizal fungi improve soil quality also by having a direct influence on soil aggregation and therefore aeration and water dynamics. An interesting potential of this fungi is its ability to allow plant access to nutrient sources which are generally unavailable to the host plants and thus plants may be able to use insoluble sources of P when inoculated with mycorrhizal fungi but not in the absence of inoculation.
- **Blue green algae (BGA):**BGA are the pioneer colonizers both in hydrosphere and xerosphere. BGA constitute the largest, most diverse and widely distributed group of prokaryotic microscopic organisms that perform oxygenic photosynthesis. These are also known as cyanophyceae and cyanobacteria. These are widely distributed in tropics; and are able to withstand extremes of temperature and drought. The significance of the abundance of BGA in Indian rice soils has been well recognized. Multi-location trials conducted under varying agro-climatic conditions have indicated that the algal inoculation could save 30 kg N/ha, however, it depends upon the agro ecological conditions. BGA has been reported to reduce the pH of soil and improve upon exchangeable calcium and water holding capacity. The recommended method of application of the algal inoculum is broadcasting on standing water about 3 to 4 days after transplantation. After the application of algal inoculum the field should be kept water logged for about a week's time. Establishment of the algal inoculum can be observed within a week of inoculation in the form of floating algal mats, more prominently seen in the afternoon.
- **Azolla***:* A floating water fern 'Azolla' hosts nitrogen fixing BGA *Anabaena azollae*. Azolla contains 3.4 per cent nitrogen (on dry wt. basis) and add organic matter in soil. This bio fertilizer is used for rice cultivation. There are six species of Azolla *viz. A. caroliniana, A. nilotica, A. mexicana, A.filiculoides, A. microphylla* and *A. pinnata*. Azolla plant has a floating, branched stem, deeply bilobed leaves and true roots which penetrate the body of water.The leaves are arranged alternately on the stem. Each

leaf has a dorsal and ventral lobe. The dorsal fleshy lobe is exposed to air and contains chlorophyll. It grows well in ditches and stagnant water. Azolla can be easily grown throughout the year in India if water is not a limiting factor and climatic conditions are favourable for its growth. This fern usually forms a green mat over water. Azolla is readily decomposed to NH_4 which is available to the rice plants. Field trial have shown that rice yields increased by 0.5-2t/ha due to Azolla application. In India and China, an about 20 and 18 per cent increase in rice yield, respectively has been reported due to Azolla application.

In organic farming, chemical herbicides cannot be used. So weeding can be done only manually. Different cultural practices like tillage, flooding, mulching can be used to manage the weeds. Besides, biological (pathogen) method can be used to manage the loss due to weeds. When the ground is fallow, a cover crop can be planted to suppress weeds and build soil quality. Weeds growth can also be limited by using drip irrigation whenever possible, which restricts the distribution of water to the plant line.

Insect Pest Management

In organic farming, the presence of pests (where and when) is anticipated in advance and accordingly the planting schedules and locations are adjusted as much as possible to avoid serious pest problems. The main strategy to combat harmful pests is to build up a population of beneficial insects, whose larvae feed off the eggs of pests. The key to building a population of beneficial insects is to establish borders (host crops) around fields planted with blends of flowering plants that the beneficial insects particularly like. Then periodically beneficial insects are released into the fields, where the host crops serve as their home base and attract more beneficial insects over time. When faced with a pest outbreak that cannot be handled by beneficial insects, the use of natural or other organically approved insecticides like neem pesticides is done. The two most important criteria for allowed organic pesticides are low toxicity to people and other animals and low persistence in the environment. These criteria are determined by the National Organic Standards.

Diseases Management in Organic Farming

Plant diseases are major constraints for reductions in crop yield and quality in organic and low input production systems. Proper fertility management to crops through balanced supply of macro and micronutrients and adoption of crop rotation have shown to improve the resistance of crops to certain diseases. Thus one of the biggest rewards of organic farming is healthy soil that is alive with beneficial organisms. These healthy microbes, fungi and bacteria keep the harmful bacteria and fungi that cause disease in check.

Organic Farming in India: Present Status and Future

India has traditionally practised organic agriculture, but the process of modernization, particularly the green revolution technologies, has led to the increased use of chemicals. In recent years, however, limitations of agriculture based on chemical use and intensive irrigation have become apparent and there has been a resurgence of interest in organic agriculture. Renewed interest in organic agriculture is mainly due to two concerns, falling agricultural yield in certain areas as a result of excessive use of chemical inputs, decreased soil fertility and environmental awareness. Exports also played a role but perhaps lesser than in other countries.

India holds a unique position among 172 countries practicing organic agriculture: it has 6, 50,000 organic producers, 699 processors, 669 exporters and 7,20,000 hectares under cultivation. But, with merely 0.4 per cent of total agricultural land under organic cultivation, the industry has a long journey ahead (Bordolo, 2016). India produced around 1.35 million MT (2015-16) of certified organic products which includes all varieties of food products namely Sugarcane, Oil Seeds, Cereals and Millets, Cotton, Pulses, Medicinal Plants, Tea, Fruits, Spices, Vegetables, Coffee *etc.* The production is not limited to the edible sector but also produces organic cotton fibre, functional food products *etc.*

Table 15.1. Percentage of Area Under Organic Farming in the Total Cultivated Area of different Countries of the World in 2004

Country	*Percentage of Area under Organic Farming*
USA	0.23
UK	4.22
Germany	4.10
Argentina	1.70
Austria	8.40
Australia	2.20
Japan	0.10
Switzerland	7.94
South Africa	0.05
Italy	3.70
India	0.03
Pakistan	0.08
Srilanka	0.05

Source: SOEL Survey (2004).

Table 15.2. The Status of Organic Production in India: 2006-2007

Total area under certified	2.8 M ha
Organic cultivation	
Total production	585970 Mt
Total quantity exported	19456 Mt
Value of total export	Rs. 30124 lakh
Number of farmers	141904

Source: GoI (2008b).

India is home to 30 per cent of the total organic producers in the world, but accounts for just 2.59 per cent (1.5 million hectares) of the total organic cultivation area of 57.8 million hectares, according to the World of Organic Agriculture 2018 report. Thus, amongst the regions with the largest areas of organically managed agricultural land, India ranked 9th.

Organic farming has the potential to change the face of agriculture in India today. Currently, we have 1.2 million ha of land under organic production, which comprises only 0.7 per cent of total area under cultivation. This produces around 1.24 million tons of organic produce. With the increase in global health consciousness, organic food is set to knock every door and make its way in healthy kitchens worldwide

Table 15.3. Export of Organic Agricultural Commodity from India (2014-15 to 2016-17)

Organic Agricultural Commodity		*Basmati Rice*	*Non-Basmati Rice*	*Other Cereals*	*Fruits and Vegetables*	*Pulses*	*Processed Items*	*Fruits/ Seeds*
2014-15	Qty (MT)	3702284	8225564	6425297	3212091	222104	721622	12498
	Rs. (Cr.)	27598.71	20336.00	10233.02	12434.95	1218.10	12195.93	124.98
2015-16	Qty (MT)	4044833	6374172	1522707	2982038	255602	728224	10684
	Rs. (Cr.)	22714.37	15085.38	2561.24	12719.60	1655.44	12738.80	106.84
2016-17	Qty (MT)	3999722	6813397	1000640	5155810	137177	1320527	11680
	Rs. (Cr.)	21605.13	17121.08	1868.49	16138.49	1281.63	13121.44	116.80

Source: APEDA

The renewed interest in organic farming in India is mainly due to three main reasons, reduction in agricultural yield in certain areas as a result of excessive and indiscriminate use of chemical inputs, decreased soil fertility and concern regarding the environment. The 10th Five-Year Plan encouraged the promotion of organic farming using organic wastes, and integrated pest management (IPM) and integrated nutrient management (INM) practices (GOI, 2001). Even the 9th Five-Year Plan had emphasized the promotion of organic produce in plantation crops, spices and condiments using organic and bio-inputs for the protection of environment and promotion of sustainable agriculture

Currently, India ranks 33rd in terms of total land under organic cultivation and 88th in agricultural land under organic crops to a total farming area. According to the Agricultural and Processed Food Product Export Development Authority (APEDA), the cultivated land under certification is around 2.8 Million hectares (2007-08), which includes one million hectares under cultivation and the rest is under forest area (wild collection). An estimated 69 Million hectares, however, is traditionally cultivated without using chemical fertilizers and could be eligible for certification under the current practices, or with small modifications. Certifying these farms remains a challenge, however, as many of these farms are small holdings (nearly 60 per cent of all farms in India are less than one ha). Smallholders and resource-poor farmers may not be able to afford the cost of certification, they are illiterate and unable to maintain necessary records or maybe using indigenous cultivation systems not recognized in organic certification systems. These farms mainly produce for home consumption and supply to the local markets in case of irregular surpluses. Such barriers pose difficulties for farms to reap the potential benefits of organic certification (Reddy, 2010).

In order to satisfy the increasing demand of organic products the organic farming sector in India has substantially increased over the course of year.

India ranks 33rd in world in terms of area under organic farming.

India rank 88th in terms of the ratio of agriculture land under organic crop to total farming area.

Madhya Pradesh has highest area under organic farming (1.1 mha or 52 per cent)

Maharashtra is at second (0.96 mha or 33.6 per cent)

Odisha is at third (0.67 mha or 9.7 per cent)

Uttrakhand and Sikkim is recognized as the organic states

Reasons for Organic Farming

The population of the planet is skyrocketing and providing food for the world is becoming extremely difficult. The need of the hour is sustainable cultivation and production of food for all.

The Green Revolution and its chemical-based technology are losing its appeal as dividends are falling and returns are unsustainable. Pollution and climate change are other negative externalities caused by the use of fossil fuel based chemicals.

In spite of our diet choices, organic food is the best choice you'll ever make, and this means embracing organic farming methods. Here are the reasons why we need to take up organic farming methods:

1. To Accrue the Benefits of Nutrients

Foods from organic farms are loaded with nutrients such as vitamins, enzymes, minerals and other micro-nutrients compared to those from conventional farms. This is because organic farms are managed and nourished using sustainable practices. In fact, some past researchers collected and tested vegetables, fruits, and grains from both organic farms and conventional farms.

The conclusion was that food items from organic farms had way more nutrients than those sourced from commercial or conventional farms. The study went further to substantiate that five servings of these fruits and vegetables from organic farms offered sufficient allowance of vitamin C. However, the same quantity of fruits and vegetables did not offer the same sufficient allowance.

2. Stay Away From GMOs

Statistics show that genetically modified foods (GMOs) are contaminating natural foods sources at real scary pace, manifesting grave effects beyond our comprehension. What makes them a great threat is they are not even labelled. So, sticking to organic foods sourced from veritable sources is the only way to mitigate these grave effects of GMOs.

3. Natural and Better Taste

Those that have tasted organically farmed foods would attest to the fact that they have a natural and better taste. The natural and superior taste stems from the well balanced and nourished soil. Organic farmers always prioritize quality over quantity.

4. Direct Support to Farming

Purchasing food items from organic farmers is a sure-fire investment in a cost-effective future. Conventional farming methods have enjoyed great subsidies and tax cuts from most governments over the past years. This has led to the proliferation of commercially produced foods that have increased dangerous diseases like cancer. It's time governments invested in organic farming technologies to mitigates these problems and secure the future. It all starts with you buying food items from known organic sources.

5. Conserve Agricultural Diversity

These days, it normal to hear news about extinct species and this should be a major concern. In the last century alone, it is approximated that 75 per cent of the agricultural diversity of crops has been wiped out. Slanting towards one form of farming is a recipe for disaster in the future. A classic example is a potato. There were different varieties available in the marketplace. Today, only one species of potato dominate. This is a dangerous situation because if pests knock out the remaining potato specie available today, we will not have potatoes anymore. This is why we need organic farming methods that produce disease and pest-resistant crops to guarantee a sustainable future.

6. To Prevent Antibiotics, Drugs, and Hormones in Animal Products

Commercial dairy and meat are highly susceptible to contamination by dangerous substances. A statistic in an American journal revealed that over 90 per cent of chemicals the population consumes emanate from meat tissue and dairy products.

According to a report by Environmental Protection Agency (EPA), a vast majority of pesticides are consumed by the population stem from poultry, meat, eggs, fish and dairy product since animals and birds that produce these products sit on top of the food chain.

This means they are fed foods loaded with chemicals and toxins. Drugs, antibiotics, and growth hormones are also injected into these animals and so, are directly transferred to meat and dairy products. Hormone supplementation fed to farmed fish and dairy products contributes mightily to the ingestion of chemicals. These chemicals only come with a lot of complications like genetic problems, cancer risks, growth of tumour and other complications at the outset of puberty.

Limitations and Implications of Organic farming

There are a few limitations with organic farming such as

1. Organic manure is not abundantly available and on plant nutrient basis it may be more expensive than chemical fertilizers if organic inputs are purchased.
2. Production in organic farming declines especially during first few years, so the farmer should be given premium prices for organic produce.
3. The guidelines for organic production, processing, transportation and certification *etc.* are beyond the understanding of ordinary Indian farmer.
4. Marketing of organic produce is also not properly streamlined. There are a number of farms in India which have either never been chemically managed/cultivated or have converted back to organic farming because of farmers' beliefs or purely for reason of economics. These thousands of farmers cultivating million acres of land are not classified as organic though they are. Their produce either sells in the open market along with conventionally grown produce at the same price or sells purely on goodwill and trust as organic through select outlets and regular specialized markets. These farmers may never opt for certification because of the costs involved as well as the extensive documentation that is required by certifiers.

Government Initiatives to Promote Organic Farming

Programmes	*Highlights*
National Project on Organic Farming (NPOF)	Central sector scheme, implemented during the 10th FYP with an outlay of Rs. 57.04 crores. Expanded in the 11th FYP with an outlay of Rs. 101 crores. Objective: to encourage the organic food production and promote manufacture and usage of organic and biological inputs, such as organic manure, bio-control agents, and biologically produced fertilizers and pesticides. Provides financial aid up to Rs. 63 lakhs (33 per cent of the capital cost) for constructing fruits and vegetables waste compost units, through NABARD. Provides subsidy up to 40 lakhs (25 per cent of the capital cost) for the construction of bio fertilizeror bio-pesticide production unit, through NABARD.
National Project on Management of Soil Health and Fertility (NPMSF)	Implemented during the 11th FYP with an outlay of Rs. 429.85 crores. Objective: to promote the judicious and balanced use of fertilizers and organic manure on the basis of soil test results. Provides financial assistance of Rs. 500/hectare for promoting the use of organic manure
Network Project on Organic Farming	Initiated by ICAR in the 10th FYP at the Project Directorate for Farming system Research at Modipuram. Objective: to develop package of practises for different crops and farming systems for organic farming in different agro-climatic conditions of India. Package of practises have been developed for: basmati rice, rainfed wheat, maize, red gram, chickpea, soybean, groundnut, mustard, isabgol, black pepper, ginger, tomato, cabbage and cauliflower

Programmes	*Highlights*
National Horticulture Mission (NHM) and Horticulture Mission for North East and Himalayan State	Centrally Sponsored Scheme, launched in 2005-06. Aim: to strengthen the growth of the horticulture sector comprising of fruits, vegetables, roots and tuber crops, mushroom, spices, flowers, aromatic plants, cashew and cocoa. Provides subsidy of 50 per cent for establishing vermin-compost units and HDPE vermin-beds. Provides assistance for organic certification of Rs. 5 lakh for a group of farmers covering an area of 50 hectares. Provides Rs. 30,000 per beneficiary for adopting organic farming.
Rashtriya Krishi Vikas Yojna (RKVY)	Provides assistance to the projects formulated and approved by the state for decentralized production and marketing of organic fertilizers.
National Mission for Sustainable Agriculture (NMSA)	100 per cent assistance by the state government for setting up of mechanization of fruit/vegetable waste. 100 per cent aid for setting up of quality control laboratory for testing bio-fertilizers, up to Rs. 85 lakhs
Paramparagat Krishi Vikas Yojana (PKVY)	Provides Rs. 20,000 the farmers up to 3 years for performing organic farming. Procuring packaging material, preparation of labels, holograms, printing and branding of organic produce at Rs. 2,500/acre Provides financial aid for a cluster of 50 acres, to the tune of Rs. 1, 20,000 for transporting organic produce to the market place. In order to motivate and support marketing facilities, financial assistance of Rs. 36,330/cluster is provided to organize an organic fair.

Major Researchable Issues in Organic Farming

- Site specific sustainable organic package of practices for major and underutilized crops of the State is required with special reference to the nutrient budgeting in crops and cropping system and changes in soil health due to use of organic nutrient sources.
- Locally available organic sources of nutrients should be identifiedand nutrients supply through the organic sources and demand of crops should be synchronized. With regards to agro-waste management's research should be focused on recycling of farm waste means in-situ fertility management, conservation agriculture *etc.*
- Isolation, identification and characterization of suitable microbes for efficient nutrient use to fulfil the nutrient demand and for controlling the plant diseases.
- Development of profitable and suitable integrated farming system models for small and marginal farmers to improve their livelihoods is the need of hour.
- Identification of most profitable cropping systems and after identification the nutrient demand of these cropping systems should be standardized for efficient use of natural resources.

- Organic seed production of different crops should be under taken under the technical guidance of plant breeder and agronomist to fulfil the seed requirement of the region.
- Carbon sequestration capacity of crops and cropping system should also be worked out to known the carbon dynamics in the soil.
- Date of sowing/planting of major crops should be re–evaluated and screening of heat/cold tolerances verities of crops especially rice and maize is required under changing clime conditions.

Suggestions and Recommendation

- The farmers' should be made aware with the scientific information about organic agriculture.
- Government should provide subsidies in organic produce to the farmers and facility of easy credit with lower rate of interest.
- Higher prices should be determined by the government for organic produce than the conventional produce.
- Agriculture universities should encourage the research in the field of organic farming.
- Government, NGO's and extension workers should organize various workshops, seminars, conferences, *etc.* with the help of subject matter specialist for farmers.
- Private companies should invest in the project of producing organic food products free from harmful chemicals.
- At an individual level, should promote the use of organic produce by going for organic agriculture in their kitchen garden, buying organic products available in the market.

Conclusion

Organic agriculture is a holistic food production system works with the sustainable use of locally available natural resources. The need is to adopt a comprehensive approach for the promotion of organic agriculture by taking cooperation of all stakeholders, environmental friendly technologies, marketing infrastructure and financial support for quality and quantity organic food production. An environmentally sustainable system of agriculture like organic agriculture will be able to maintain a resource balance, avoid over exploitation of resource, conserving soil nutritional quality, health and biodiversity.

REFERENCES

Bordolo, B. (2016). The future lies in organic farming. The Hindu Business Line.

De, L.C. Organic Farming, Principles and Practices.

Deshpande, R.S. (2002). Suicides by farmers in Karnataka: Agrarian distress and possible alleviatory steps. *Economic and Political Weekly*, XXXVII (26): 2601-2610.

FAO (1999). Organic Agriculture, Food and Agriculture Organization of the United Nations, Rome.

GoI (Government of India) (2001). Report of Working Group on Organic Farming and Biodynamic Farming for the 10th Five-Year Plan, Planning Commission, New Delhi, September

Ram, B. (2003). Impact of human activities on land use changes in arid Rajasthan: Retrospect and prospects. In: *Human Impact on Desert Environments*, Eds: P. Narain, S. Kathaju, A. Kar, M.P. Singh and Praveen Kumar, Scientific Publishers, Jodhpur, pp. 44-59.

Reddy Suresh, B. (2010a). Soil fertility managment in semiarid regions: The socio-cultural, economic and livelihood dimensions of farmers' practices — A case of Andhra Pradesh, unpublished PhD Thesis, Centre for Economic and Social Studies, Dr. B.R. Amedkar University, Hyderabad

Reddy Suresh, B. (2010b). Assessment of Economic and Ecological Returns from Millet- based Bio-diverse Organic Farms vis-à-vis Conventional Farms, CESS Monograph Series No.8, Centre for Economic and Social Studies, Hyderabad

The Indian Organic Market: A New Paradigm in Agriculture, ASSOCHAM, India, 2018.

Chapter 16

Cryogenic Freezing: A Concise Review

Fozia Hameed[1], Neeraj Gupta[2], Monika Sood[2], Julie D. Bandral[3] and Rukhsana Rahman[1]

[1]Ph.D Scholar, [2]Asst. Professor, [3]Assoc. Professor
Division of Food Science and Technology, S.K. University of Agricultural Sciences and Technology of Jammu, Chatha – 180 009, J&K

Since time immemorial freezing is considered one of the most versatile and convenient techniques for long-term food preservation while maintaining most of the fresh-like characteristics of food and food products. Historical evidence suggests that natural food freezing using winter frost, ice cellars, and compressed snow in insulated cellars have been employed for food preservation from as early as 1000 BC. Freezing techniques have evolved at a rapid pace with key improvements to process efficiency, productivity, cost reduction and enhancement of food product quality with extended shelf life. Developments in freezing technology evolved rapidly with changes in consumers lifestyles coupled with the advancement in freezing technologies. However, quality control and food preservation by conventional freezing is not a recent phenomenon. Further, the increasing trend of transporting fresh foods in the international market has prompted researchers to search for newer freezing technologies. Maintenance of proper quality of frozen food demands thorough knowledge of cryobiology, which is considered as one of the recent developments in freezing technology. It has now been recognized that the ultra fast freezing methods are technically possible by cryogenic means for all food materials and are also economically viable for a wide variety of products. Cryogenic technology gives low temperature application in food industry. This technology controls ice nucleation in such way that water can transform into small ice crystals. Fast freezing rates result in the formation of large number of small ice crystals, leading to a higher quality of frozen products as compared to slower freezing methods. Further, cryogenic freezing is better suited to small or medium-sized products or products that have a large surface area. So, cryofreezing is the only viable method available for long-term preservation of both plant and animal origin products.

***Keywords**: Cryofreezing, Ice crystals, Quality, Shelf life, Food preservation.*

Freezing is considered as a superior processing technique with the potential to deliver improved microbial safety, nutritional profile, organoleptic quality, and convenience. Freezing preservation has distinct advantages over other conventional food preservation techniques including canning and drying in terms of nutritional properties and quality. During freezing, water undergoes phase transition at subzero temperatures leading to ice crystal formation, retarding biochemical changes and inhibiting the growth of microorganisms and the activity of enzymes responsible for food spoilage (Ojha *et al.*, 2016). However, tissue damage in frozen foods inevitably occurs due to the differential density of ice and water. The physically damaged tissue liberates moisture (drip loss) upon thawing, resulting in the loss of its nutritional quality. Discerning consumers increasingly require that the quality of frozen food products must be comparable to that of fresh food. Thus the frozen food industry must ensure that the original product freshness, color, and flavor, including nutritional qualities, are not significantly damaged during the freezing process. To minimize this damage, rapid freezing techniques are strongly recommended (Hong *et al.*, 2005). Ultra rapid freezing, the best of all freezing technique is well known. Cryogen can meet up this requirement easily. Now-a-days, researchers are also trying to use cryogen for production of dairy products such as ice cream. Transportation of frozen foods can be effectively carried using cryogens. Several methods have been evolved and are real substitute of mechanical refrigeration techniques on board. Solid carbon-dioxide is used for transporting heat sensitive food materials commercially. If properly designed and planned, cryogens can be friendly to the food industry.

Cryogenic Freezing

Cryogenic freezing of food has been applied in the food industry since the early 1960s and is well suited for many of the product categories such as frozen processed red meat, poultry, and fish/seafood, frozen processed fruit and vegetables, frozen ready-to-eat meals, pizza, and meat substitutes, frozen bakery products, frozen processed potatoes (including uncooked and oven baked), frozen dessert products (including ice cream) (MarketLine, 2014). The use of expandable gaseous refrigerants such as argon, hydrogen, nitrogen, oxygen, carbon dioxide, and other gases, which sublime or evaporate at very low temperatures at atmospheric pressure, is referred to as cryogenic freezing. In food manufacturing, the most commonly used cryogenic substances are nitrogen (N_2) and carbon dioxide (CO_2) (Estrada-Flores, 2016). The boiling-off of the refrigerant when it comes in contact with the product brings about cooling. As well as using the latent heat absorbed by the boiling liquid, sensible heat is absorbed by the resulting cold gas. Most cryogenic systems use total loss refrigerants, *i.e.*, the refrigerant is released to the atmosphere and not recovered. Due to environmental and economic factors, total loss refrigerants must be both readily available and harmless, which limits the choice to atmospheric air and its components. During a typical cryogenic process, the surface of the product can be exposed to, a spray of liquid N_2; a mixture of solid ('snow') and gaseous CO_2; or a direct immersion into the liquid cryogen (Almqvist, 2003).

History

Cryogenic freezing was first carried out commercially using liquid air, in the 1930s. However, liquid air contains a high proportion of liquid oxygen, which is a powerful oxidizing agent. Theoretically it can be produced on site, eliminating the need to purchase and store 'gas.' Although companies have promoted the use of liquid air, in practice it has been superseded by less harmful liquid nitrogen and liquid, or solid, carbon dioxide.

Food freezing using liquid nitrogen or carbon dioxide became established in Europe from the early 1960s onwards as manufacturers looked for uses of these byproducts (Fennama, 1978). Liquid nitrogen is produced by liquefaction of air either as a principle product or byproduct of liquid oxygen, while carbon dioxide is produced during fermentation processes and is a constituent of natural air. By the early 1990s an estimated 10 per cent of all frozen food in the UK was produced using cryogens. In UK about 100,000 tons of nitrogen was used per year by the food industry for freezing or transporting of goods.

In 1968 by Du Pont, liquid dichlorodifluoromethane (food grade R 12, freon, CCl_2F_2), was introduced commercially as an alternative to the air-based total loss cryogens. R 12 had the advantage over other cryogens of being easily recoverable and having a higher operating temperature (-29.8 °C), reducing the problems of thermal shock. It was claimed that such gentle handling allows 'true' individual quick freezing (IQF) of products such as soft berries, meat patties, raw shrimps, breaded scampi, fried onion rings, *etc.* within 6 min or less. However, its use was not permitted in all countries and it was not widely used (James and James, 2003).

Properties of Cryogens

Cryogens are the ultra cold fluids. Direct immersion can be inadvisable since an extremely high temperature gradient can be imposed on the food, which is often sufficient to cause it to disintegrate. Research studies have concluded that the ideal cryogen would have a boiling point of -50 °C and a latent heat of evaporation as high as possible (Table 16.1).

Table 16.1. Physical and Chemical Properties of Nitrogen and Carbon Dioxide

Chemical Formula	N_2	CO_2
Molecular weight (kg/mol)	28	44
Latent heat (kJ/kg)	199.1 (vaporization)	572.3 (sublimation)
Density of vapor at 0 °C (kg/m^3)	1.26	1.97
Processing temperature (°C)	-195.5 (boiling point)	-78.3 (solid)

Liquid nitrogen is produced by liquefaction of air, either as a principal product or as a byproduct of liquid oxygen. Nitrogen is the main constituent of atmospheric air and at atmospheric pressure liquefies at a temperature of -196 °C. It is usually supplied and stored at a pressure of 3–6 bar, with corresponding boiling points of -185 °C to -177 °C (Lang, 2006). A useful rule of thumbis that 1th^{-1} of liquid nitrogen is approximately equivalent to 100 kW of mechanical refrigeration.

Since liquid nitrogen is made from air, its costs are entirely those of manufacture, *i.e.*, the capital cost of the plant, electricity, and transportation. The energy required to produce 1 kg of liquid nitrogen is around 3000 kJ, or eight times the consequent stored refrigerating effect.

Carbon dioxide's physical properties are unusual, in that it does not exist in liquid form at atmospheric pressure. If stored as a pressurized liquid and released into the atmosphere, the liquid changes partly to gas and partly to a frozen solid at -78.5 °C, which sublimes directly into gas without going through a liquid phase.

Liquid carbon dioxide is generally supplied either at ambient temperature (*e.g.*, 25 °C and 65 bar), giving a refrigerating capacity of 199 kJ kg^{-1}, or at -16 °C and 22 bar, giving a refrigerating capacity of 311 kJ kg^{-1}. At the point of use, spray nozzles reduce the pressure of the liquid, generating a mixture of cold vapor and solid carbon dioxide 'snow.' The majority of the 'refrigeration effect' stored in solid CO_2 is latent, whereas in liquid nitrogen almost half the effect is due to sensible heat transfer to the cold gas (Pham *et al.*, 2006).

Heat Transfer during Cryogenic Freezing

In general, the rate of heat transfer during freezing is influenced by several factors, including:

- The product's thermal (*e.g.*, thermal conductivity, specific heat) and physical (*e.g.*, surface area available for heat transfer, size, shape) properties.
- The temperature difference between the food and the freezing medium
- The presence of thermal barriers such as packaging materials or air gaps within the food structure.

The most common process in cryogenic freezing consists of spraying the surface of the product with either N_2 or CO_2. When liquid CO_2 is fed into a spray nozzle, the CO_2 expands and changes to approximately equal parts (by weight) of solid and vapor. As solid CO_2 particles contact the food surface, sublimation occurs almost instantly, drawing latent heat out of the product. This heat transfer mechanism provides ~85 per cent of the total cooling required to freeze the product (Pham, 2014). The remaining cooling is a result of different heat transfer mechanisms that include the convective currents created by the resulting CO_2 'snow' and the gaseous CO_2 traveling through the tunnel, the radiative heat transfer between the product and the surrounding (colder) surfaces of the equipment, and the conductive heat from the product to the belt. The importance of each of these mechanisms varies according to the specific product characteristics (shape in particular) and design features of the equipment.

In case of freezers that use liquid N_2, the liquid is sprayed into the freezer and separates into liquid droplets and vapor. As droplets touch the product surface, the liquid N_2 becomes gaseous N_2, extracting latent heat from the food surface in this process. Cryogenic freezing with N_2 draws about 50 per cent of the refrigeration effect through the N_2 phase change (OSHA, 2015). The rest of the product cooling and freezing occurs via convective heat transfer between the gaseous N_2 flowing along the freezer and in countercurrent with the product flow, the radiative heat transfer between the product and the surrounding (colder) surfaces of the equipment, and the conductive heat from the product to the belt. In case of tunnel freezers, the significance of these heat transfer mechanisms on spiral freezing depends on the specific product characteristics and design features of the equipment.

Mechanism

During cryogenic immersion freezing, heat removal is a boundary phenomenon mostly dictated by the temperature difference between the product's surface and the liquid cryogen. The latter boils when it contacts the surface of the product, leading to very high heat transfer coefficients (Pham, 2014). For liquid N_2, the surface heat transfer coefficient should theoretically vary from 200 W/m^2K in the film boiling region to about 8000 W/m^2K (Singh and Mannapperuma, 1990). However, there is a scarcity of systematic experiments in the scientific literature that provide accurately measured heat transfer parameters for cryogenic freezing systems. Table 16.2 summarizes some published heat transfer coefficients that were either measured or calculated.

Table 16.2. Heat Transfer Coefficients during Cryogenic Operations

Operation	*Heat Transfer Coefficient ($Wm^{-2}K^{-1}$)*	*Source*
Immersion of hamburgers in liquid N_2	170	Agnelli and Mascheroni (2001)
Immersion of meatballs in liquid N_2	235	Agnelli and Mascheroni (2001)
Immersion of strawberries in liquid N_2	190	Agnelli and Mascheroni (2001)
Cooling of eggs in an experimental setup simulating a cryogenic CO_2 tunnel	10 (during cooling) 12 (during equilibration)	Sabliov *et al.*(2002)
Tunnel freezing of raspberries and blueberries with liquid N_2	25 (precooling section) 175 (freezing section) 60 (subcooling section)	Zosifescu *et al.* (2013)

The use of accurate heat transfer parameters for the calculation of freezing times in cryogenic systems is as important as the selection of a method to calculate these processing times. Finite difference (Shaikh and Prabhu, 2007) and finite element methods (Zosifescu *et al.*, 2013) are often selected to model cryogenic freezing.

Pham's method (1984) is an approximate analytical method that models the freezing process as the sum of three stages: precooling, phase change, and subcooling. The precooling and subcooling times are calculated using a modification to Newton's cooling law which takes into account the effect of internal resistance. The phase change time is calculated according to Plank's equation. By applying simple correction factors, Pham's method provided freezing times within ±10 per cent of the results provided by the finite difference method.

Equipment

Tunnel Freezers

Essentially a tunnel freezer is a countercurrent heat exchanger. The cryogen is sprayed at one end and the gas passed countercurrent to product traveling on a belt in the opposite direction. This enables the exit temperature to be only 20 °C below the product inlet temperature. Thus product can be precooled before being subjected to the full force of the refrigeration effect. Approximately 50 per cent of the product's heat can be extracted before it reaches the cryogen spray. Straight or spiral tunnels are utilized. A tunnel freezer operates by conveying the product through an enclosed space within which the cryogen is discharged. The product passes through three distinct zones in the tunnel: the precool zone, the freezing zone, and the postcool zone. In the precool zone, the product is cooled using gas and fans force gas vertically past the product. In the freezing zone, sprays discharge cryogen directly on product and belt, while in the postcool zone, which is a gas-cooled section, with more fan circulation, some temperature stabilization takes place (Estrada-Flores, 2016).

Immersion Freezers

Immersion freezers are the simplest and most inexpensive way of using cryogens to freeze foodstuffs. The principle of these systems is the immersion of the product in a liquid N_2 bath, freezing the surface of the product almost instantly. In immersion freezing, the main objective is to achieve a superficial 'crust,' reducing dehydration and product clumping. Immersion freezers are best suited to high volumes of product where crust freezing is required or where the product is sufficiently robust to withstand the thermal shock of a high temperature differential (The Linde Group, 2015). Very fast heat transfers are possible, but the system is wasteful in that only the latent heat of the liquid is used and cold vapor is lost.

Cabinet Freezers

Batch cabinet freezers utilizing cryogens have been developed for freezing prepared and precooked food, particularly for airline meals. Liquid cryogen is injected at time intervals into a cabinet containing the food on racks. Circulation fans are usually utilized to provide even and efficient freezing. Freezing times vary from 5 min to 1 h, depending on the product and its temperature (James and James, 2003).

Cryogenic Impingement Freezers

These systems use an initial stage of liquid N_2 spray to produce a superficial crust in the product, followed by the application of a high-speed gaseous cryogen flow perpendicular to the surface of the product. The use of jets drastically decreases the superficial resistance to heat transfer and thus, the freezing rate relies mostly on the temperature gradient between product and N_2, which is typically of 190 °C, and the thermal properties of the product. Impingement freezers are generally better suited to products with high surface-to-weight ratios (Newman and McCormick, 2001). Although impingement freezers can raise the production rates in 20–50 per cent above the capacities obtained with traditional cryogenic tunnels, the latter are 15–40 per cent less costly than impingement freezers (Lang, 2006).

Combined Freezers

A cost-effective alternative to 100 per cent cryogenic freezing is a combined freezer. The principle is to use a cryogenic freezing unit for initial freezing of the outer surface of the product, followed immediately by mechanical freezing to reduce the temperature of the bulk of the product. One advantage of this type of approach is that the cryogenic stage can often be retrofitted (Gazda and Koziol, 2013).

Effect of Cryogenic Freezing on Quality of Frozen Products

The debate about the relationship between freezing rate and quality has gone on for many years. Many trade publications indicate that a considerable improvement in quality can be obtained by ultrarapid freezing with cryogens. Some work has indicated that there is an interaction between freezing rate and cooking method. When meat was cooked after thawing, no differences were found between freezing rates (13, 2, or 0.04 cm h^{-1}) in terms of flavor, juiciness, tenderness, thaw loss, or TBA (thiobarbituric acid) values. However, meat that had been cooked from frozen was found to show a favorable effect of faster freezing rates.

With mushrooms, while the quantity of the drip loss is almost independent of the freezing rate, the color of the lost juice changes conspicuously. It contains more coloring and flavoring ingredients the slower the freezing. Among the products that obtain a decisive and lasting quality improvement by rapid freezing are carrots, corn on the cob, mushrooms, egg products, sauces, cream, shrimps, and prawns.

Appearance

Freezing rate does affect the appearance of many food products. In case of poultry, fast freezing tends to produce a lighter-colored product as the small ice crystals scatter the light more than larger crystals. This lighter color is preferred by many consumers. In fish, rapidly frozen fillets have a dense white, opaque appearance, in comparison with the dark, translucent, vitreous product produced by air blast freezing (Rodezno *et al.*, 2013).

Drip Loss

When ice crystals grow they puncture the cell walls resulting in juice bleeding. This causes 'drip loss' when the product is thawing. The larger the average crystal size is, the greater the number of punctured cells and the greater the drip loss. The loss of juice can result in a loss of firmness and flavor. The drip loss from some food (*i.e.*, blueberries) can be clearly related to freezing rate (Alfaro *et al.*, 2018). Slow freezing can produce drip losses of 20 per cent. Fast freezing without surface cracking can reduce losses to approximately 5 per cent. With meat and meat products freezing considerably increases drip.

Dehydration–Weight Loss

During freezing, product dehydration is the result of several phenomena, including an increase in solute concentrations in the unfrozen medium as a result of the formation of ice, the water vapor pressure difference between the product surface and surrounding storage environment, and structural damage to cells leading to rupture and loss of intracellular water (Goff, 2002). During cryogenic freezing lower dehydration occurs, so lower weight loss, primarily due to the rapid cooling and the fast decrease of water vapor pressure at the surface. Weight loss during freezing of beefburgers can be as high as 3 per cent in an air blast freezing system compared with 0.4 per cent in a liquid nitrogen tunnel. In a modern spiral freezer operating at -30 °C air temperature, the weight loss will be approximately 1.2 per cent. In an air impingement freezer operating at a -46 °C air temperature, the weight loss from a burger can be <0.4 per cent (Estrada-Flores, 2016).

Freeze-Cracking

Products like delicate soft fruits and light pastry products, may crack when they are submitted to very high freezing rates or very low temperatures. Research suggests that crust freezing produces a shell that prevents further volume expansion, when the internal portion of the unfrozen material undergoes phase transition. If the internal stress is higher than the frozen material strength, the product will crack during freezing (James and James, 2003). Products with high void spaces, which allow internal stresses to dissipate, show a lessened chance of freeze-cracking. Precooling prevents freeze-cracking because it reduces the differences in temperature between the product and the freezing medium, and reduces the difference between freezing time at the center and at the surface. When the phase change of the core region occurs before the surface becomes brittle, food products can withstand the internal pressure and thus freeze-cracking does not occur (Pham *et al.*, 2006).

Microbiological Activity

As, freezing prolongs the shelf life of products by decreasing the rate of microbial growth. However the negative effect of freezing on microorganisms relates to temperature shock, concentration of extracellular solutes, toxicity of intracellular solutes, dehydration and ice formation (Zaritzky, 2000). Cryogenic freezing is one of the preferred methods for maintaining viable organs, tissues, and cells during long-term storage, although consumers and manufacturers should not take microbiological safety for granted in cryogenically frozen products (Everstine *et al.*, 2013).

Considering the above, cryogenic freezing has been proven to successfully reduce microbiological risks in some cases. One example is the use of cryogenic treatments to reduce the presence of *Vibrio vulnificus* in oysters. After 35 days of cryogenic freezing, *Vibrio* presence in oysters was shown to decrease to nondetectable levels (Mestey and Rodrick, 2006).

Advantages of Cryogenic Freezing

Advantages that can accrue from cryogenic freezing, at least in some instances, are as follows:

1. **Good retention of original quality attributes.** In most instances, the quality of cryogenically frozen foods (*e.g.*, carrots) equals or exceeds that of conventionally frozen foods, provided that the period of frozen storage is limited to a few weeks or less and the temperature of frozen storage is -18 °C or lower. The rapid freezing achieved with cryogenic refrigerants results in numerous small ice crystals that are uniformly distributed throughout the sample, minimal dislocation of water, a near-normal cellular structure in the frozen state, and minimal time in the temperature zone (0 to 5 °C below the initial freezing point of the product) where the bulk of the ice forms and physical and chemical damage to the product is most likely (Xu *et al.*, 2015).

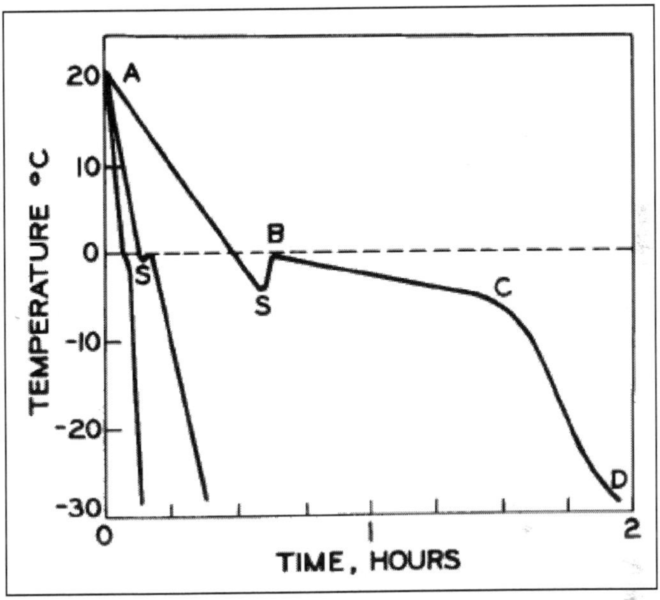

Figure 16.1. Schematic Freezing Curves Typical of Foods Frozen at different Rates. B is the initial freezing point, S is the point of maximum supercooling, and BC is the zone during which most of the water is converted to ice.

Cryogenic freezing, in case of poultry, has an additional advantage in that it imparts a desirable chalky-white appearance to the surface of the frozen product. However, the same effect can be achieved by freezing rapidly by other procedures (Alhamdan *et al.*, 2015).

2. **Minimal loss of product weight during freezing:** Loss of water during freezing from the product can be a significant economic factor, particularly with products of high value. Freezing of unpackaged food in air-blast freezers can result in weight losses in excess of 4 per cent, however properly operated cryogenic freezers can limit weight losses to 0.5 per cent or less, which is generally better than well-operated air-blast or plate freezers (Rodezno *et al.*, 2013). Though, it should be noted that if a liquid

nitrogen freezer is operated to consume a minimal amount of liquid nitrogen per pound of product (maximum use of nitrogen vapors to precool the incoming product; final product temperature of -10 °C instead of - 20°C), weight losses from the product can exceed 1 per cent.

3 **Low capital investment**: It is generally recognized that cryogenic freezers that utilize liquid nitrogen or liquid carbon dioxide involve capital investments of less than one-half that required for air-blast or plate freezers of the same capacity (Harr and Minard, 1977). This advantage is lost if the cryogenic system involves liquid carbon dioxide with a recovery system or Freezant 12.

4. **Minimal space required**: Cryogenic freezers generally occupy considerably less space than an air-blast or plate freezer of the same capacity.

5. **Versatility**: Cryogenic freezers, particularly those operating with liquid nitrogen or liquid carbon dioxide are adaptable to a wide variety of product shapes and sizes. Furthermore, these types of freezers are especially advantageous when the production rate is variable or of uncertain duration.

However, there are also some disadvantages associated with cryogenic freezing, such as higher operating costs as compared to mechanical refrigeration systems (Chourot *et al.*, 2003). Further, cryogenic freezing is better suited to small or medium sized products, or products that have a large surface area. In large products with limited surface area, the rate of freezing – which is a key advantage in cryogenic freezing – is limited by the rate of internal heat transfer (Cleland and Valentas, 1997). Cryogenic freezing suits small-scale production, new or seasonal products that require fast freezing, and situations where the installed freezing capacity cannot cope with production throughput demands (ASHRAE, 2006).

Conclusion

Cryogenic systems can be used effectively to freeze, transport, and store a wide range of foodstuffs, and a wide range of equipment and techniques are available using either nitrogen or carbon dioxide. Cryogenics is involved in over 10 per cent of the frozen food market and a minute proportion of the chilling market. Scientific evidences so for are sufficient to prove that cryogenic freezing can effectively prolong the shelf life of fresh fruits and vegetables without affecting nutritional properties as compared to conventional freezing techniques. Microprocessors combined with freezing programs give the technology of cryofreezing a new lead. This will eventually enable frozen and thawed products to be saleable as fresh.

REFERENCES

Agnelli, M. E. and Mascheroni, R. H. (2001). Cryomechanical freezing. A model for the heat transfer process. *J. Food Eng.* 263–270.

Alhamdan, A., Hassan, B., Alkahtani, H., Abdelkarim, D. and Younis, M. (2015). Cryogenic freezing of fresh date fruits for quality preservation during frozen storage. *Journal of the Saudi Society of Agricultural Sciences*, 1-8.

Alfaro, L., Siramard, S., Chouljenko, a. and Sathivel, S. (2018). Effects of liquid nitrogen pretreatment on the osmotic dehydration and quality of cryogenically frozen blueberries (*Vaccinium angustifolium* Ait.). *Food Bioscience,* **42** (16): 1-18.

Almqvist, E. (2003). History of Industrial Gases. Kluwer Academic/Plenum Publishers, New York.

Ashrae. (2006). Industrial Food Freezing Systems. *ASHRAE Handbook:* Refrigeration. The American Society of Heating, Refrigerating and Air-C.

Chourot, J. M., Macchi, H., Fournaison, L. and Guilpart, J. (2003). Technical and economical model for the freezing cost comparison of immersion, cryomechanical and air blast freezing processes. *Energy Convers. Manage,* **44,** 559–571.

Cleland, D. J. and Valentas, K. J. (1997). Prediction of freezing time and design of food. *Handbook of Food Engineering.*

Estrada- Flores, S. 2016. Cryogenic freezing of foods. Plant and Food Research, Melbourne, Victora, Australia, 1-9.

Everstine, K., Spink, J., Kennedy, S. (2013). Economically motivated adulteration (EMA) of food: common characteristics of EMA incidents. *J. Food Protect.,* **76,** 723–735.

Fennama, O. (1978). Cryogenic freezing of foods. *Advances in cryogenic engineering,* 712-720.

Gazda, W. and Koziol, J. (2013). The estimation of energy efficiency for hybrid refrigeration system. *Appl. Energy,* 101: 49–57.

Goff, G. H. (2002). Theoretical Aspects of the Freezing Process. *University of Guelph,*

Harr, J. W. and Minard, M. E. (1977). Proceedings of the Meat Industry Research Conference. *American Meat Institute,* 25-33.

Hong, G. P., Park, S. H., Kim, J. Y., Lee, C. H., Lee, S., and Min, S. G. (2005). The effect of thawing rate on the physicochemical properties of frozen ostrich meat. *Food Sci. Biotechnol.,* **14**: 676-680.

James, C. and James, S. (2003). Cryogenic freezing. *Encyclopedia.* 2725-2732.

Lang, G., 2006. Cryogenic Freezing. Industrial Refrigeration Consortium Research and Technology Forum. University of Wisconsin-Madison, Madison, WI.MarketLine, 2014. Global Frozen Food. Informa.

Mestey, D. and Rodrick, G. E. (2006). A comparison of cryogenic freezing techniques and their usefulness in reduction of *Vibrio vulnificus* in whole oysters. *Int. Proc. Molluscan Shellfish Saf.* 467–474.

Newman, M. D. and McCormick, S. 2001. Modular Apparatus for Cooling and Freezing of Food Product on a Moving Substrate. USA Patent Application 6263680.

OSHA, 2015. Occupational Health and Safety Guideline for Nitrogen [Online]. US Dept. of Labor. Available: https: //www.osha.gov/dts/chemicalsampling/data/CH_257300.html (accessed 01.10.15).

Ojha, K. S., Kerry, J. P., Tiwari, B. K. and O'Donnell, C. (2016). Freezing for Food Preservation. *Elsevier*, 1-9.

Pham, Q. T. (1984). Extension to Planck's equation for predicting freezing times of foodstuffs of simple shapes. *Int. J. Refrig.*, **7**, 377–383.

Pham, Q. T., Le Bail, A. and Tremeac, B. (2006). Analysis of stresses during the freezing of solid spherical foods. *Int. J. Refrig.*, 125–133.

Pham, T. Q. (2014). Freezing time formulas for foods with low moisture content, low freezing point and for cryogenic freezing. *J. Food Eng.*, **127**, 85–92.

Sabliov, C. M., Farkas, B. E., Keener, K. M. and Curtis, P. A. (2002). Cooling of shell eggs with cryogenic carbon dioxide: a finite element analysis of heat.

Shaikh, N. I. and Prabhu, V. (2007). Mathematical modeling and simulation of cryogenic tunnel freezers. *J. Food Eng.*, **80**, 701–710.

Singh, R. P. and Mannapperuma, J. (1990). Developments in food freezing. *Bioprocess and Food Process Engineering*, Marcel Dekker, Inc., New York, NY.

Rodezno, L. A. E., Sundararajan, S., Solval, K. M., Chotiko, A., Li, J., Alfaro, L., Bankston, J. D. and Sathivel, S. (2013). Cryogenic and air blast freezing techniques and their effect on the quality of catfish fillets. *LWT - Food Science and Technology*, **54:** 377-382.

The Linde Group (2015). Cryoline®SI Immersion IQF Freezer.

Xu, Congcong., Wang, Liping., Shao, Liyu., Yu, H. and Li, Y. (2015). Effect of freezing/thawing temperature on the viscoelastic and nutritional qualities of carrots. *International Journal of Food Properties*, 1-37.

Zaritzky, N. E. (2000). Factors affecting the stability of frozen foods. *Managing Frozen Foods*, Woodhead Publishing Ltd.

Zosifescu, C., Damian, V. and Coman, G. (2013). Cryogenic freezing of berries-assessment of phase change heat transfer. *ARA Annual Congress Proceedings*, 454–457.

Chapter 17

Application of Nanoscience in Rhizosphere Studies and Smart Delivery Systems

Shubham Singh, Sanjay Swami, Rakesh Maurya and Pritisha Patgiri*

School of Natural Resource Management, College of Post Graduate Studies in Agricultural Sciences, CAU (I), Umiam, Meghalaya

Agriculture is the backbone of most developing countries, with more than 60 per cent of the population reliant on it for their livelihood. Agricultural scientists are facing a wide spectrum of challenges such as stagnation in crop yields, low nutrient use efficiency, declining soil organic matter, multi- nutrient deficiencies, climate change, shrinking arable land and water availability and shortage of labour besides exodus of people from farming. In spite of immense constraints faced, we need to attain a sustainable growth in agriculture at the rate of 4 per cent to meet the food security challenges. To address these problems, there is a need to explore one of the frontier technologies such as 'Nanotechnology' to precisely detect and deliver the correct quantity of nutrients and pesticides that promote productivity while ensuring environmental safety and higher use efficiency. The nanotechnology can be exploited in the value chain of entire agriculture production system. Nanotechnology is emerging as the sixth revolutionary technology in the current era after the Industrial Revolution of Mid 1700s, Nuclear Energy Revolution of the 1940s, The Green Revolution of 1960s, Information Technology Revolution of 1980s and Biotechnology Revolution of the 1990s. It is now emerging and fast growing field of science which is being exploited over a wide range of disciplines such as physics, chemistry, biology, material science, electronics, medicine, energy, environment and health sectors. Nanoscience has brought revolution in different fields by helping develop processes and products that are hardly possible to evolve through conventional methods. The nanotechnology aided applications have the potential to change agricultural production by allowing better management and conservation of inputs of plant and animal production. Nanomaterials hold great promise regarding their application in agriculture in terms of plant protection and nutrition due to their size-dependent qualities, high surface-to-volume ratio and unique optical properties. Agriculture nanotechnology used under the entire field to enhancing the productivity with using smart delivery system of fertilizer, insecticide, herbicide and growth regulator. The related fields are as under food technology, crop improvement, seed technology, precision farming, plant disease diagnoses, weed management, water management, biosensors and pest management. The world demand for fertilizer is increasing day by day. Therefore, there is an urgent need to tackle the excessive usage of pesticides and fertilizers. Nanotechnology provides solutions to a complex set of engineering and scientific challenges in the vital food industry manufacturing safe food of high quality. Nanoscience principles can apply in the field of soil science as nano-fertilizer, biosensors and in rhizosphere studies for improving the fertilizer capacity and its management. However, this technology may also be associated with several risks in the environmental and health fields and all safety issues and the negative impacts of micro-particles must be studied before their release into the environment. So, it's very important to know the importance of nanotechnology for present and future use.

***Keywords**: Nanotechnology, Biosensors, Rhizosphere studies, Nano-fertilizers.*

Agriculture is the backbone of most developing countries, with more than 60 per cent of the population reliant on it for their livelihood. Agricultural scientists are facing a wide spectrum of challenges such as stagnation in crop yields, low nutrient use efficiency, declining soil organic matter, multi- nutrient deficiencies, climate change, shrinking arable land and water availability and shortage of labour besides exodus of people from farming. In spite of immense constraints faced, we need to attain a sustainable growth in agriculture at the rate of 4 per cent to meet the food security challenges. To address these problems, there is a need to explore one of the frontier technologies such as 'Nanotechnology' to precisely detect and deliver the correct quantity of nutrients and pesticides that promote productivity while ensuring environmental safety and higher use efficiency. The nanotechnology can be exploited in the value chain of entire agriculture production system (Subramanian and Tarafdar, 2011). Nanotechnology is emerging as the sixth revolutionary technology in the current era after the Industrial Revolution of Mid 1700s, Nuclear Energy Revolution of the 1940s, The Green Revolution of 1960s, Information Technology Revolution of 1980s and Biotechnology Revolution of the 1990s. It is now emerging and fast growing field of science which is being exploited over a wide range of disciplines such as physics, chemistry, biology, material science, electronics, medicine, energy, environment and health sectors.

Nantechnology deals with the matter at nanoscale (1-100 nm) dimensions. These materials when reduced to the nanoscale show some properties which are different from what they exhibit on a macro scale, enabling unique applications. Nanoscience has brought revolution in different fields by helping develop processes and products that are hardly possible to evolve through conventional methods. The nanotechnology aided applications have the potential to change agricultural production by allowing better management and conservation of inputs of plant and animal production. A survey by Salamanca–Buentella *et al.* (2005) predicted several nanotechnology applications for agricultural production for developing countries within next 10 years. These included -

1. Nanoforms zeolites for slow release and efficient dosage of water and fertilizers for plants; drugs for livestock; nanocapsules and herbicide delivery.
2. Nanosensors for soil quality and for plant health monitoring; nanosensors for pest detection.
3. Nanomagnets for removal of soil contaminants and
4. Nanoparticles for new pesticides, insecticides, and insect repellents.

The recent statistics suggest that about 90 per cent of the nano-based patents and products have come from just seven countries, namely, China, Germany, France, Japan, Switzerland, South Korea and USA while India's investments and progress is far from satisfactory. However, to take advantage of the fascinating field of nano-science, the Government of India has invested Rs. 1000 crore through the Nano Mission Project during 11th Five Year Plan and the investment is likely to be several folds higher during the 12th Five Year Plan period.

The Indian Council of Agricultural Research (ICAR) has opened up an exclusive platform to target Nanotechnology Applications in Agriculture. The ICAR – Nanotechnology Platform encompasses major themes such as synthesis of nano-particles for agricultural use, quick diagnostic kits for early detection of pests and diseases, nano-pheromones for effective pest control, nanoagri- inputs for enhanced use efficiencies, precision water management, stabilization of organic matter in soil, nano food systems and bio safety besides establishing the policy frame work. Green-synthesis and microbial synthesis of nanomaterials for their agricultural use may be very important as they are naturally encapsulated with mother protein, therefore, more stable and safer to biological system. At present in India research is mainly concentrated on nano particle synthesis, smart release of nutrients from nano- fertilizers, nano-induced polysaccharide powder for moisture retention/soil aggregation and C build up, regulated release of active ingredients from nano-encapsulated herbicides, nano-seed invigoration, and slow and steady release of pesticides, nano- film for extended shelf- life of perishables and nano-remediation of soil and aquatic pollutants. These are cutting-edge researchable areas which are expected to expand in the years to come. However, if the nanoproducts and the processes for creating them are not managed judiciously, there could be serious health and environmental risks.

Application of Nanotechnology in Agriculture and Allied Science

Agriculture is the backbone of most developing countries, with more than 60 per cent of the population reliant on it for their livelihood (Brock *et al.*, 2011). As well as developing improved systems for monitoring environmental conditions and delivering nutrients or pesticides as appropriate, nanotechnology can improve our understanding of the biology of different crops and thus potentially enhance yields or nutritional values. In addition, it can offer routes to added value crops or environmental remediation. Nanotechnology, as a new enabling technology, has the potential to revolutionize agriculture and food systems. Agricultural and food systems security, disease treatment delivery systems, new tools for molecular and cellular biology, new materials for pathogen detection and protection of the environment are examples of the important links of nanotechnology to the science and engineering of agriculture and food systems (Welch and Graham, 1999). Nanotechnology operates at the same scale as a virus or disease infecting particle, and thus holds the potential for very early detection and eradication. Nanotechnology holds out the possibility that "Smart" treatment delivery systems could be activated long before macro symptoms appear. For example, a smart treatment delivery system could be a miniature device implanted in an animal that samples saliva on a regular basis. Long before a fever develops, the integrated sensing, monitoring and controlling system could detect the presence of disease and notify the farmer and activate bioactive systems such as drugs, pesticides, nutrients, probiotics, nutraceuticals and implantable cell bioreactors.

Nanofertilizer

Fertilizer play pivotal role in the agriculture production up to 35 to 40 per cent of the productivity. To enhance nutrient use efficiency and overcome the chronic problem of eutrophication, nanofertilizer might be a best alternative. Attempts have been made to synthesize nanofertilizer in order to regulate the release of nutrients depending on the requirements of the crops, and it is also reported that nanofertilizer are more efficient than ordinary fertilizer.

Nanofood

The definition of nanofood is that nanotechnology techniques or tools are used during cultivation, production, processing or packaging of the food. It does not mean atomically modified food or food produced by nanomachines. Although there are ambitious thoughts of creating molecular food using nanomachines, this is unrealistic in the foreseeable future. Instead nanotechnologists are more optimistic about the potential to change the existing system of food processing and to ensure the safety of food products, creating a healthy food culture. They are also hopeful of enhancing the nutritional quality of food through selected additives and improvements to the way the body digests and absorbs food. Although some of these goals are further away, the food packaging industry already incorporates nanotechnology in products. In Agriculture nanotechnology used under the entire field to enhancing the productivity with using smart delivery system of fertilizer, insecticide, herbicide and growth regulator.

Precision Farming Smart Field System

- Bio-Nanotechnology has designed sensors which give increased sensitivity and earlier response to environmental changes and linked into GPS
- These monitor soil conditions and crop growth over vast areas
- Such sensors have already been employed in US and Australia UASD Plant pathology 100 per cent growth inhibition was seen in the *Pythium ultimum, Magna porthegrisea, Colletotrichum gloeosporioides, Botrytis cinere* and *Rhyzoctonia solani,* showed at 10 ppm of the nanosized silica-silver (Park *et al.,* 2006).

- ☆ QDs have emerged as pivotal tool for detection of a particular biological marker with extreme accuracy (Madhuri *et al.*, 2010).

Soil Remediation

- ☆ Nanotech-based soil binder called Soil Set employed to avoid soil erosion Sequoia Pacific Research of Utah (USA).

Removal of Heavy Metals

- ☆ Ligand based nanocoating can be utilized for effective removal of heavy metals as these have high absorption tendency.

Nano-fertilizer Technology

In India, fertilizers, alongwith quality seed and irrigation, are mainly responsible for enhanced food grain production (55 mt) in 1960s to (254 mt) in 2011 coinciding with the spectacular increase in fertilizer consumptions from 0.5 mt to 23 mt, respectively. It has been conclusively demonstrated that fertilizer contributes to the tune of 35-40 per cent of the productivity of any crop. Considering its importance, the Government of India is heavily subsidising the cost of fertilizers particularly urea. This has resulted in imbalanced fertilization and occurrence in some areas, nitrate pollution of ground waters due to excessive nitrogen application. In the past few decades, use efficiencies of N, P and K fertilizers have remained constant as 30-35 per cent, 18-20 per cent and 35-40 per cent, respectively, leaving a major portion of added fertilizers to accumulate in the soil or enter into aquatic system causing eutrophication. In order to address issues of low fertilizer use efficiency, imbalanced fertilization, multi-nutrient deficiencies and decline of soil organic matter, it is important to evolve a nano-based fertilizer formulation with multiple functions.

Nano-fertilizer technology is very innovative but scantily reported in the literature. However, some of the reports and patents strongly suggest that there is a vast scope for the formulation of nano- fertilizers. Significant increase in yields have been observed due to foliar application of nano particles as fertilizer (Tarafdar, 2012; Tarafdar *et al.*, 2012a). It was shown that 640 mg ha^{-1} foliar application (40 ppm concentration) of nanophosphorus gave 80 kg ha^{-1} P equivalent yield of clusterbean and pearl millet under arid environment. Currently, research is underway to develop nano-composites to supply all the required essential nutrients in suitable proportion through smart delivery system. Preliminary results suggest that balanced fertilization may be achieved through nanotechnology (Tarafdar *et al.*, 2012b). Indeed the metabolic assimilation within the plant biomass of the metals, *e.g.*, micronutrients, applied as Nano- formulations through soil-borne and foliar application or otherwise needs to be ascertained. Further, the Nano-composites being contemplated to supply all the nutrients in right proportion through the "Smart" delivery systems also needs to be examined closely. Currently, the nitrogen use efficiency is low due to the loss of 50-70 per cent of the nitrogen supplied in conventional fertilizers. New nutrient delivery systems that exploit the porous nanoscale parts of plants could reduce nitrogen loss by increasing plant uptake. Fertilizers encapsulated in nanoparticles will increase the uptake of nutrients (Tarafdar *et al.*, 2012c). In the next generation of nanofertilizers, the release of the nutrients can be triggered by an environmental condition or simply released at desired specific time. Foliar application of nano Phosphorous as fertilizer (640 mg ha-1) and soil application of phosphorous fertilizer (80 kg ha^{-1}) yielded equally in clusterbean and pearl millet under arid environment (Tarafdar *et al.*, 2012)

Nano-pesticide

Persistence of pesticides in the initial stage of crop growth helps in bringing down the pest population below the economic threshold level and to have an effective control for a longer period. Hence, the use of active ingredients in the applied surface remains one of the most cost-effective and versatile means of controlling insect pests. In order to protect the active ingredient from the adverse environmental conditions and to promote persistence, a nanotechnology approach, namely "nano-encapsulation" can be used to improve the insecticidal value. Nano encapsulation comprises nano-sized particles of the active ingredients being sealed by a thin- walled sac or shell (protective coating). Recently, several research papers have been published on the encapsulation of insecticides. Nano-encapsulation of insecticides, fungicides or nematicides will help in producing a formulation which offers effective control of pests while preventing accumulation of residues in soil. In order to protect the active ingredient from degradation and to increase persistence, a nanotechnology approach of "controlled release of the active ingredient" may be used to improve effectiveness of the formulation that may greatly decrease amount of pesticide input and associated environmental hazards. Nano-pesticides will reduce the rate of application because the quantity of product actually being effective is at least 10-15 times smaller than that applied with classical formulations, hence a much smaller than the normal amount could be required to have much better and prolonged management. Several pesticide manufacturers are developing pesticides encapsulated in nanoparticles. These pesticides may be time released or released upon the occurrence of an environmental trigger (for example, temperature, humidity, light). It is unclear whether these pesticide products will be commercially available in the short-term. Plant diseases are major factors limiting crop yields. The problem with the disease management lies with the detection of the exact stage of prevention. Most of the time appropriate plant protection chemicals are applied to the crop as a precautionary measure leading to avoidable environmental hazards, or else applications are made after the appearance of the disease symptoms, thereby causing some amount of crop losses. Among the different diseases, the viral diseases are the most difficult to control, as one has to stop the spread of the disease by the vectors. But once it starts showing its symptoms, pesticide application would not be of much use. Therefore, detection of the exact stage such as stage of viral DNA replication or the production of initial viral protein is the key to the success of control of viral diseases. Nano-based viral diagnostics, including multiplexed diagnostics kits development, have taken momentum in order to detect the exact strain of virus and the stage of application of some therapeutic to stop the disease. Detection and utilization of biomarkers, that accurately indicate disease stages, is also an emerging area of research in bio-Nanotechnology. Measuring differential protein production in both healthy and diseased states leads to the identification of the development of several proteins during the infection cycle. Clay nanotubes (halloysite) have been developed as carriers of pesticides at low cost, for extended release and better contact with plants, and they will reduce the amount of pesticides by 70-80 per cent, thereby reducing the cost of pesticide with minimum impact on water streams.

Nano-scale Carriers

Nanoscale carriers can be utilized for the efficient delivery of fertilizers, pesticides, herbicides, plant growth regulators, *etc.* The mechanisms involved in the efficient delivery, better storage and controlled release include: encapsulation and entrapment, polymers and dendrimers, surface ionic and weak bond attachments among others.

These help to improve stability against degradation in the environment and ultimately reduce the amount to be applied, which reduces chemical runoff and alleviates environmental problems. These carriers can be designed in such a way that they can anchor plant roots to the surrounding soil constituents and organic matter. This can only be possible if we unrevel the molecular and conformational mechanisms between the nanoscale delivery and targeted structures, and soil fractions. Such advances as and when they happen will help in slowing the uptake of active ingredients, thereby reducing the amount of inputs to be used and also the waste produced.

Biosensors to Detect Nutrients and Contaminants

Protection of the soil health and the environment requires the rapid, sensitive detection of pollutants and pathogens with molecular precision. Soil fertility evaluation is being carried out for the past sixty years with the same set of protocols which may be obsolete for the current production systems and in the context of precision farming approaches. Accurate sensors are needed for in situ detection, as miniaturized portable devices, and as remote sensors, for the real-time monitoring of large areas in the field.

These instruments are able to reduce the time required for lengthy microbial testing and immune assays. Application of these instruments include detection of contaminants in different bodies such as water supplies, raw food materials and food products. Enzymes can act as a sensing element as these are very specific in attachment to certain biomolecules. Electronic nose (E-nose) is used to identify different types of odour; it uses a pattern of response across an array of gas sensors. It can identify the odorant, estimate the concentration of the odorant and find characteristic properties of the odour in the same way as might be perceived by the human nose. It mainly consists of gas sensors which are composed of nanoparticles *e.g.* ZnO nanowires. Their resistance changes with the passage of a certain gas and generates a change in electrical signal that forms the fingerprint pattern for gas detection. Biosensors provide high performance capabilities for use in detecting contaminants in food or environmental media. They offer high specificity and sensitivity, rapid response, user-friendly operation, and compact size at a low cost (Amine *et al.*, 2006). While the direct enzyme inhibition sensors currently lack the analytical ability to discriminate between multiple toxic substances in a sample (such as simultaneous presence of heavy metal and pesticide), they may prove useful as a screening tool to determine when a sample contains one or more contaminants. These methods are amenable to deployment in single-use test strips (making them useful to those in the field). According to Hu *et al.* (2010), detection of multiple residues of organophosphorus pesticides has been accomplished using a Nano magnetic particle in an enzyme linked immunosorbent assay (ELISA) test. The authors suggest that ELISA is more cost-effective than analytical tests requiring expensive laboratory equipment with high levels of skill.

Application of Nanotechnology in Rhizosphere Studies

Rhizosphere is a narrow region just adjacent the root system that is influenced by rootexudates and associated microbes. This region is highly biological active region having a higher bacterial concentrations compare to surrounding bulk soil. The plant and rhizosphere microorganisms communicate with each other through 'signals'. Root microbiota contributes 40 per cent ofthe plant's photosynthesis fixation of carbon into the atmosphere.Synergistic interactions between bacteria and root surfaces that occur in the rhizosphere are a critical part of the nitrogen and carbon cycles, and are essential for plant growth. These synergistic interactions between bacteria and roots in the rhizosphere region will affect nutrient exchange process and hinder plant growth if affected by external means. Several rhizosphere microorganisms are referred to as plant growth–promotingrhizobacteria (PGPR) and fungi (PGPF), which can improve plant growth and health.

Microorganisms form significant interactions with a widearray of biota in terrestrial ecosystems, often with important consequences for all species present. Terrestrial plant species are colonized by a large number of microorganisms in several morphological regions, including interior tissues (endophytes), on their leaves (epiphytes) and in their rhizosphere.Rhizosphere microbiomeplays a major mechanism in reprogramming the defence responses in host plant. Microbe-mediated biological control is one of the potentialoptions in plant disease management.Different plant host system in the same soil may select different microbialcommunities, most probably because of variations in root exudates. In the rhizosphere, the microbial genes greatly outnumber the plant genes, resulting in a microbiome that has been referred to as the plant's second genome. Just like the human gut microbiome, the plant's rhizosphere microbiome has been implicated in many functions that support its host's well-being. For example, rhizosphere microorganisms can increase nutrient uptake, protect against pathogens, enhance abiotic stress tolerance and promote the development of the plant. However, some rhizosphere organisms are also capable of causing plant disease or impeding plant growth. Alternatively, plants also influence the microbial community within their rhizosphere, largely mediated through root structure and exudates that determine the physical and chemical conditions within the plant root zone. As root structure and exudates change during the plant's life cycle, theconditions in the rhizosphere are altered, likely selecting for adifferent microbial community. Other factors such as xenobiotics can disturb the rhizosphere conditions, having both direct and indirectconsequences for the plant and its microbial community.

Conclusion

Recently studies on substantial microbial diversity in the rhizosphere werepossible by the sequencing technologies. These studies confirm the rhizospheremicroorganisms consists of a subset of the total diversity in the soil mass, whereinplants are developed. Soil nature is a significant feature in determining rhizosphere microbial groupand it acts as a reservoir for the selection of rhizosphere inhabitants. Even though soilis the decisive feature in organization of root-associated microbial communities, thereis overlay in the microbial species, which are genotypically selected in related plantswith diagonally different soils. This recommends the existence of co-evolutionary relationships betweenplant-inhabiting organisms and their hosts.Plant-microbe interactions are favourable feature to enhance the plant health. It is evident that, root microbiome providesessential strength to the plant. Plantscan particularly select and enrich few species by the secretion of compounds thatselectively stimulate or suppress microbial growth. Thedisease suppressive ability in the soils is usually rooted in specific microbialpopulations that inhibit the pathogens. Disease-suppressive soils exist in all-inclusiveareas and some of them expanded by subsequent prolonged monoculture of a specificcrop. Disease suppressive ability of soils iscontributed by various microorganisms' likes, Trichoderma, Fusarium, Streptomyces,Bacillus, and Actinomycetes spp. Pseudomonas spp. is recognized as the mostimportant organism. Potential actions of disease suppressioninclude competition for space and micronutrients like, antagonism, hyper parasitismwith microbial production of secondary metabolites, such as iron-chelatingsiderophores, antibiotics, and lytic enzymes; and activation of ISR.

REFERENCES

Amine, A., Mohammadi, H., Bourais, I. and Palleschi, G. (2006). Enzyme inhibitionbased biosensors for food safety and environmental monitoring. *Biosensors and Bioelectronics*, 21: 1405-1423.

Astruc, D., Boisselier, E. and Ornelas, C. (2010). Dendrimers designed for functions: from physical, photophysical, and supramolecular properties to applications in sensing, catalysis, molecular electronics, and nanomedicine. *Chemical Review*, 110: 1857-1959.

Brock, D.A., Douglas, T.E., Queller, D.C. and Strassmann, J.E. (2011). Primitive agriculture in a social amoeba. *Nature*, 469: 393-396.

Chinnamuthu, C. R. and Kokiladevi, E. (2007). Weed management through nanoherbicides. In: Application of nanotechnology in agriculture. C.R. Chinnamuthu, B. Chandrasekaran, and C. Ramasamy (Eds.) Tamil Nadu Agricultural University, Coimbatore, India.

Dreizin, E.L. (2009). Metal-based reactive nanomaterials. *Prog. Energy Combustion Science*, 35: 141-167.

Hu, Y., Shen, G., Zhu, H. and Jiang, G. (2010). A class-specific enzyme linked immunosorbent assay based on magnetic particles for multi-residue organophosphorus pesticides. *Journal of Agricultural and Food Chemistry*, 58: 2801–2806.

Khodakovskaya, M., Dervishi, E., Mahmood, M., Yang, Xu.,Zhongrui, Li., Watanabe, F. and Biris, A.S. (2009). Carbon nanotubes are able to penetrate plant seed coat and dramatically affect seed germination and plant growth. ACS Nano, 3: 3221–3227

Madhuri, S., Choudhary, A.K. and Kumar, R. (2010). Nanotechnology in agricultural diseases and food safety. *Journal of Phytopathology*, 2: 83 – 92.

Oberdorster, E., Zhu, S., Blickley, T.M., Green, P.M. and Haasch, M.L. (2006). Ecotoxicology of carbonbased engineered nanoparticles: Effects of fullerene (C60) on aquatic organisms. *Carbon*, 44: 1112-1120.

Park, H.J., Kim, S.H., Kim, H.J. and Choi, S.H. (2006). A new composition of nanosized silica-silver for control of various plant diseases. *Plant Pathology Journal*. 22: 295–302.

Petit, C., Lixon, P. and Pileni, M.P. (1993). In situ syntheses of silver nanocluster in AOT reverse micelles,*Journal of Physical Chemistry*, 97: 12974-12983.

Roco, M.C. (2003). Broader Societal Issues of Nanotechnology. *Journal of Nanoparticle Research*, 5: 181-189.

Salamanca-Buentello, F., Persad, D. L., Court, E. B., Martin, D. K., Daar, A. S. and Singer, P. A. (2005). Nanotechnology and the developing world. *PLoS Medicine*, 2: 383–386.

Sastry, R.K., Rashmi, H.B. and Rao, N.H. (2011). Nanotechnology for enhancing food security in India. *Food Policy*, 36: 391–400.

Subramanian, K. S. and Tarafdar, J. C. (2011). Prospects of nanotechnology in Indian farming. *Indian Journal of Agricultural Sciences*, 81: 887-893.

Tarafdar, J.C. (2012). Perspectives of nanotechnological applications for crop production. *AAS News*, 12: 8-11.

Tarafdar, J.C., Agrawal, A., Raliya, R., Kumar, P., Burman, U. and Kaul, R.K. (2012b). ZnO nanoparticles induced synthesis of polysaccharides and phosphatases by Aspergillus fungi. *Advanced Science, Engineering and Medicine*, 4: 1-5.

Tarafdar, J.C., Raliya, R. and Rathore, I. (2012a). Microbial synthesis of phosphorus nanoparticles from Tri-calcium phosphate using *Aspergillus tubingensis* TFR-5. *Journal of Bionanoscience*, 6: 84-89.

Tarafdar, J.C., Xiang, Y., Wang, W.N., Dong, Q. and Biswas, P. (2012c). Standardization of size, shape and concentration of nanoparticle for plant application. *Applied Biological Research*, 14: 138-144.

Verma, A. K., Singh, V. P. and Vikas, P. (2012). Application of nanotechnology as a tool in animal products processing and marketing: an overview. *American Journal of Food and Technology*, 10: 389-398.

Welch, R.M. and Graham, R.D. (1999). A New Paradigm for World Agriculture: Meeting Human Needs, Productive, Sustainable, and Nutritious. *Field Crops Research*, 60: 1-10.

Chapter 18

A Case Study of Farmers about Knowledge of Pea Production Technology in Hadoti Region of Rajasthan

N.R. Meena[1] and R.K. Dohrey[2]*

[1]Assistant Professor, [2]Professor,
College of Agriculture, ANDUAT, Kumarganj, Ayodhya, Uttar Pradesh

The present study was conducted on knowledge of farmers regarding modern cultivation practices of pea production in Hadoti (Bundi, Kota and Tonk districts) region of Rajasthan. Two tehsils from each identified districts were selected on the basis of maximum area under pea cultivation. Thus, in all six tehsils were taken for the present study. Total twenty villages were identified on the basis of proportionate sampling from the selected tehsils. The total 200 respondents were selected from above mention 3 districts for collection of information regarding knowledge of pea production. Further, it was reported from the study that out of 200 respondents, 50.50 per cent respondents had medium level of knowledge, while 28.50 and 21.00 per cent farmers having high and low knowledge level regarding pea production technology respectively. It was also noted that extent of knowledge of marginal farmers was from 38.73 to 80.60 per cent, whereas in case of small farmers the knowledge was found to be from 40.20 to 82.30 per cent in all improved cultivation practices of pea. There was significant difference in knowledge between marginal and small farmers about improved pea production technology.

Keywords: *Pea production, Knowledge, Technology, Farmers, Rajasthan.*

Pea (*Pisum sativum* L.) is one of the important vegetable crops grown all over the world. It is a cool season crop of the tropical and sub-tropical regions. The important pea producing countries are USA, China, India, France, U.K., Netherlands, Hungry, Russia, Egypt and Australia. Archaeological evidences suggest the existence of pea (*Pisum sativum* L.) in India before the advent of the Aryans. It was used in pre-historic period during Indus valley civilization (C. 2500-300 B.C.) as evidenced by the carbonized pea seeds discovered in Mohan-ga-Daro and Harappa excavations. The sweet garden pea (*Pisum sativum* Var. hortense) varieties developed in temperate regions of Europe might have been introduced in India during the British regime. Some of the hardy and round seeded garden pea varieties like Bonia, Kaip, Asauji, *etc.* may possibly be the early introductions of Britishers. These were less sensitive to wilt even under warmer conditions and have been used in different breeding programme.

In India, pea occupies on area of 370.0 thousand hectares with the production of 3517.0 thousand mt. (Indian Horticulture Database-2011). Pea is grown as winter vegetable in the plains of north India and as summer vegetable in the hills. It is cultivated in Uttar Pradesh, Madhya Pradesh, Himachal Pradesh, Punjab, Haryana, Rajasthan, Maharashtra, Bihar and Karnataka, which contributing to 67 per cent of the total production. Pea is a cool season crop best suited to production in the temperate regions. In the warmer areas of the tropics pea is restricted to production in the cooler highlands. Temperature between 7° and 24°C are suitable for plant growth but optimum yield are achieved between 13° and 21°C. It is relatively tolerant to frost when compared with solanaceous and cucurbitaceous crops. Blossoms and pods are susceptible to frost, whereas leaves and stems are relatively tolerant. Seed germinates better at the soil temperature of 10-18.3°C. The pea crop is produced primarily under rainfed conditions but can also be irrigated. Irrigations most common for the fresh market crop. Pea is the major vegetable crop of rabi season in Kota region of Rajasthan. The productivity is low of this crop as compared to recommended by the scientists. Looking to the above facts the present study aimed to find out the knowledge of farmers about improved pea cultivation technology.

Materials and Methods

The present study was conducted in Kota region of Rajasthan. Kota region consist of five districts, out of which three districts namely Bundi, Kota and Tonk were selected purposively on the basis of maximum area under pea cultivation. Two tehsils from each identified districts were selected on the basis of maximum area under pea cultivation. Thus, in all six tehsils were taken for the present study. Total twenty villages were identified on the basis of proportionate sampling from the selected tehsils. To select the respondents, a comprehensive list of all pea growers was prepared for all villages. Thereafter, the farmers were categorized into two groups *i.e.* small and marginal on the basis of pea cultivation. The respondents selected randomly from each category of the farmers. It was planned to select 10 respondents *i.e.* five in each category from the each selected village. Thus, the total sample size of the study was 200 respondents. Data were collected through pre-structured interview schedule. Thereafter, data were analysed, tabulated and interpretated in the light of the objective.

Results and Discussion

Distribution of Farmers on the Basis their Existing Knowledge about Improved Pea Production Technology

To distribute the farmers on the basis of their existing knowledge level about pea production technology, three categories *viz.* low, medium and high level of knowledge were made by using mean score and standard deviation of the obtained knowledge scores by the respondents.

Table 18.1 depicts that out of 200 respondents, 50.50 per cent respondents had medium level of knowledge about improved pea production technology. This was followed by 21.00 per cent farmers having low knowledge level regarding pea production technology. Whereas, only 28.50 per cent respondents were observed in the high knowledge group. Analysis of table further reveals that 24.00 and 18.00 per cent marginal farmers and small farmers respectively were in the low knowledge group. Likewise, 50.00 and 51.00 per cent marginal and small farmers had medium level of knowledge respectively. Whereas, 26.00 per cent marginal farmers and 31.00 per cent small farmers possessed high knowledge about improved pea cultivation technology. The knowledge level of small farmers was slightly higher due to more size of land holding, more income level, active social participation, more cosmopolitan in nature and more prone to change than marginal farmers.

Table 18.1. Distribution of Respondents on the Basis of their Knowledge Level Regarding Improved Pea Production Technology

n = 200

Sl. No.	*Level of knowledge*	*Marginal Farmers*		*Small Farmers*		*Total*	
		Frequency	*Per cent*	*Frequency*	*Per cent*	*Frequency*	*Per cent*
1.	Low (< 47)	24	24.00	18	18.00	42	21.00
2.	Medium (47–74)	50	50.00	51	51.00	101	50.50
3.	High (> 74)	26	26.00	31	31.00	57	28.50
	Total	**100**	**100.00**	**100**	**100.00**	**200**	**100.00**

The present findings are in accordance with the findings of Angadi *et al.* (1992) who revealed that majority of farmers (58 per cent) possessed medium knowledge of cultivation practices of groundnut. Dongardive (2002) also observed that majority (71.33 per cent) of the chilli growers had low level of knowledge, while 15.33 and 12.34 per cent respondents had medium and high level of knowledge respectively about improved chilli cultivation.

Aspect-Wise Extent of Knowledge of Farmers about Improved Pea Production Technology

The data accorded in Table 18.2 indicate that the extent of knowledge about soil and field preparation aspect for pea cultivation was highest among small farmers with 82.30 per cent, while in case of marginal farmers the extent of knowledge was 79.20 per cent about soil and field preparation aspect and ranked second by the marginal farmers. The majority of both the categories of respondents were fully acquainted with the type of soil required for pea cultivation, quantity of FYM/ha is added to the soil and number of ploughing to be done for field preparation for successful cultivation of pea.

Table 18.2. Extent of Knowledge of Farmers Regarding Improved Pea Cultivation Practices

n = 200

Sl. No.	Improved Practices	Marginal Farmers		Small Farmers		Total	
		MPS	Rank	MPS	Rank	MPS	Rank
1.	High yielding varieties	54.39	7	60.96	7	57.68	7
2.	Soil and field preparation	79.20	2	82.30	1	80.75	2
3.	Soil treatment	38.73	12	40.20	12	39.46	12
4.	Seed treatment	43.63	11	46.62	11	45.13	11
5.	Time of sowing	80.60	1	81.35	2	80.98	1
6.	Seed rate and recommended spacing	58.53	6	69.50	6	64.02	6
7.	Manures and fertilizer application	68.37	3	73.80	4	71.09	3
8.	Water management	65.35	4	72.70	5	69.03	5
9.	Weed management	61.85	5	78.23	3	70.04	4
10.	Plant protection measures	44.82	10	46.62	10	45.72	10
11.	Harvesting techniques	48.10	9	52.95	9	50.53	9
12.	Marketing	52.39	8	53.96	8	53.18	8

MPS: Mean Per cent Score.

The extent of knowledge about time of sowing was 81.35 and 80.60 per cent among small and marginal farmers. The knowledge of both the categories of farmers was more or less same about sowing time of pea in the study area. This practice was ranked first by marginal farmers and second by small farmers. Almost all the farmers possessed knowledge about second week of October to first week of November is the appropriate time of sowing for pea in Kota region of Rajasthan.

Further analysis of table shows that the extent of knowledge about manure and fertilizers application for pea cultivation was 68.37 and 73.80 MPS among marginal and small farmers respectively. It was ranked third and fourth by the marginal and small farmers respectively. It was also found that few farmers knew the actual doses and times of application of manures and fertilizers in the pea crop. The knowledge about water management among marginal and small farmers was recorded to be 65.35 and 72.30 per cent respectively. Majority of the pea growers of the area were possessed knowledge about time of flowering (after 45 days of sowing) and time of pod development stage are the critical stages of irrigation in pea crop.

The knowledge about method and quantity of using potassium chloride was poor among the pea growers. It was also noted that 61.85 and 78.23 per cent knowledge about weed management was found in marginal and small farmers respectively. Majority of the pea growers were aware with hand weeding and chemical weed control practices.

The knowledge about seed rate and recommended spacing was ranked sixth by both marginal and small farmers with 58.53 and 69.50 per cent knowledge respectively. Further analysis of data reveals that nearly fifty per cent of the respondents had exact knowledge about recommended seed rate (80-100 kg/ ha) for pea/ha. The extent of knowledge about high yielding varieties of pea among marginal and small farmers was 54.39 and 60.96 per cent respectively. The knowledge of small farmers was higher than marginal farmers about high yielding varieties of pea. This aspect was ranked seventh by both the categories of pea growers. It was also noticed that pea growers possessed knowledge about the name of high yielding varieties of pea namely T-163, RPG-3, Rachna, Pusa Mugta, Pusa Panna (ODR-27) and DMR-7, whereas they were ill awared about duration and average yield of these recommended varieties of pea.

Analysis of Table 2 further shows that the extent of knowledge about marketing of produce was 52.39 and 53.96 per cent among marginal and small farmers respectively. Some of the pea growers had good knowledge about regulated market, village wholesalers and nearby mandies where green pod of pea can be marketed for good price of pea. The knowledge about harvesting techniques of pea was 48.10 and 52.95 per cent among marginal and small farmers respectively. The knowledge about this aspect was ranked ninth by both marginal and small farmers categories. It was found that most of the respondents were harvesting pea at appropriate time with hand picking method. Further it was observed that very few respondents possessed knowledge about green pea can be preserved for future use.

The knowledge about plant protection measures was ranked tenth by both, marginal and small farmers with 44.82 and 46.62 per cent knowledge respectively. Further analysis of data reveals that most of the respondents had knowledge regarding diseases of pea namely powdery mildew, root rots, rust, mosaic but they were lacking in knowledge for their chemical control measures. It was also found that nearly 40.00 per cent pea growers possessed good knowledge of common insect pests of pea and their control measures. The extent of knowledge about seed treatment, it was recorded that marginal and small farmers had 43.63 and 46.62 per cent respectively.

Nearly forty per cent of the pea growers did not know the name and quantity of chemicals which can be used for treatment of pea seed at the time of sowing. Analysis of table also reveals that the extent of knowledge about soil treatment for pea cultivation was 38.73 and 40.20 per cent respectively. The knowledge about this aspect was ranked last by both the categories of respondents.

Thus, from the above results can be concluded that extent of knowledge of marginal farmers was from 38.73 to 80.60 per cent, whereas in case of small farmers the extent of knowledge was found to be from 40.20 to 82.30 per cent in all improved cultivation practices of pea. Further, it was inferred that small farmers had more knowledge than marginal farmers about all the pea cultivation practices.

The present findings are supported by the findings of Wankhede *et al.* (1997) who reported that majority of onion growers lacked the knowledge about the important practices such as recommended varieties, seed treatment with fungicide, fertilizer application, occurrence of pests and diseases and curing as bulb produce. The majority of them had medium level of knowledge about improved practices of onion cultivation.

Comparison of Knowledge of the Pea Growers

The comparison of knowledge about improved pea cultivation technology was made under following heads:

Comparison of Knowledge between Marginal and Small Farmers

To find out the significant difference between marginal and small farmers about knowledge of improved pea cultivation technology, 'Z' test was applied. The results of the same have been presented in Table 18.3.

Hypotheses

NH_{01}: There is no significant difference in knowledge between marginal and small farmers about improved pea production technology.

RH_1: There is significant difference in knowledge between marginal and small farmers about improved pea production technology.

Table 18.3. Comparison of Knowledge between the Marginal and Small Farmers about Improved Pea Production Technology

Sl.No.	*Category of Respondents*	*Mean*	*S.D.*	*'Z' value*
1.	Marginal farmers	68.42	6.8	5.06**
2.	Small farmers	74.40	9.7	

** Significant at 1 per cent level.

Table 18.3 reveals that calculated 'Z' value was greater than its tabulated value at 1 per cent level of significance. Thus, the null hypothesis (NH_{01}) was rejected and research hypothesis (RH_1) was accepted. It reveals that there was significant difference in knowledge between marginal and small farmers about improved pea production technology. The mean value further indicates that small farmers possessed more knowledge than marginal farmers about improved pea cultivation technology. This difference in the level of knowledge might be due to the fact that small farmers had more income, extension contacts and high participation in out reach programmes as compared to marginal farmers.

The present findings are conformity with the findings of Vashishtha (2011) who revealed that there was significant difference in knowledge between tribal and non-tribal farmers about chilli cultivation technology. She further found that non-tribal farmers had higher knowledge than tribal farmers about chilli cultivation technology.

Comparison of Knowledge between Selected Three Districts

To find out the significant difference between the farmers of selected districts analysis of variance test ('F' test) was applied. The results are presented in Table 18.4.

Hypotheses

NH_{02}: There is no significant difference in knowledge between the pea growers of selected three districts.

RH_2: There is significant difference in knowledge between the pea growers of selected three districts.

Table 18.4. Comparison of Knowledge between the Pea Growers of Selected Districts about Pea Production Technology

n = 200

Sl.No.	*Source of Variation*	*d.f.*	*S.S.*	*M.S.S.*	*'F' cal*
1.	Between the districts	2	12940.74	6470.37	14.08**
2.	Error	197	90540.26	459.59	
	Total	199	103481.00		

Mean value table

Sl.No.	*Name of District*	*Mean Value*	*C.D. Value*
1.	Bundi	74.25	1.185
2.	Kota	71.20	
3.	Tonk	68.25	

** Significant at 1 per cent level.

Table 18.4 reveals that the calculated 'F' value 14.08 is higher than the tabulated value at 1 per cent level of significance and 2 degree of freedom. Thus, the research hypothesis (RH_2) entitled "to find out significant difference in knowledge between the pea growers of selected three districts" was accepted and null hypothesis (NH_{02}) was rejected. It infers that there was significant difference in knowledge between the pea growers of selected three districts about improved pea production practices. By comparing the mean value with C.D. value, there is significant difference among the pea growers of all three districts namely Bundi, Kota and Tonk with regard to possession of knowledge about improved pea cultivation technology. Similar results were found by Meena (2001), who reported that there is significant difference in level of knowledge regarding improved production practices of pea among big, small and marginal category of farmers.

Conclusions

From the above results it can be concluded that out of 200 respondents, 50.50

per cent respondents had medium level of knowledge, while 28.50 and 21.00 per cent farmers having high and low knowledge level regarding pea production technology respectively. It was also noted that extent of knowledge of marginal farmers was from 38.73 to 80.60 per cent, whereas in case of small farmers the extent of knowledge was found to be from 40.20 to 82.30 per cent in all improved cultivation practices of pea. There was significant difference in knowledge between marginal and small farmers about improved pea production technology.

REFERENCES

Angadi, I.G., Jahagirdhar, K.A. and Shinde, P.S. (1992). Awareness and knowledge of farmers about improved cultivation practices of groundnut. *Maharashtra Journal of Extension Education* **2**: 356-357.

Dongardive, V.T. (2002). A study on adoption of recommended technology of chilli crop by chilli growers in Anand district of Gujarat state. M.Sc. (Ag.) thesis submitted to Gujarat Agricultural University, Anand.

Meena, R. (2001). Role of Krishi Vigyan Kendra in adoption of improved production practices of groundnut by the farmers of Bikaner district of Rajasthan. Ph.D. (Ag) thesis submitted to Maharana Pratap University of Agriculture and Technology, Udaipur.

Meena, R. (2001). Role of Krishi Vigyan Kendra in adoption of improved production practices of groundnut by the farmers of Bikaner district of Rajasthan. Ph.D. (Ag) thesis submitted to Maharana Pratap University of Agriculture and Technology, Udaipur.

Menariya, P.L. (2000). A study of knowledge and training needs of farmers about soyabean (*Glycine max* L.) production technology in Nimbahera tehsil of Chittorgarh district, (Raj.). M.Sc. (Ag.) Thesis submitted to Maharana Pratap University of Agriculture and Technology, Udaipur.

Mewara, R.C. and Pandya, R.D. (2007). Knowledge and adoption level of tomato growers regarding value added techniques in Navsari. *Rural India* **70**: 164-169.

Mutkule, S.R., Wattamwar, V.T., Narkar, G.S. and Geete, M.H. (2001). Constraints in adoption of chilli technology. *Maharashtra Journal of Extension Education* **20**: 85-87.

Vashishtha, U. (2011). An Assessment of Knowledge and Adoption of Chilli (*Capsicum annum* L.) Production Technology in Udaipur District of Rajasthan. Ph.D. thesis submitted to MPUAT, Udaipur (Raj.).

Wankhede, P.P., Shinde, P.S. and Bhople, S.R. (1997). Constraints in adoption of onion production technology. *Maharashtra Journal of Extension Education* **16**: 347-349.

Chapter 19

A Case Study of Processing Behaviors of Farmers in Pea Production in Hadoti Region of Rajasthan

N.R. Meena and Supriya

Assistant Professor, College of Agriculture, ANDUAT, Kumarganj, Ayodhya, Uttar Pradesh

A study was conducted in Hadoti region ((Bundi, Kota and Tonk districts) of Rajasthan to measure the processing behavior of pea growers. The findings show that majority of the farmers (45.50 per cent respondents) had low information processing behavior and 31.50 per cent farmers having medium level of information processing behavior. Further, revealed that 51.00 and 40.00 per cent marginal and small farmers respectively were in the low information processing behavior group. Likewise, 28.00 and 35.00 per cent marginal and small farmers had medium information processing behavior about improved pea cultivation practices respectively. Whereas, 21.00 per cent marginal farmers and 25.00 per cent small farmers were found in high information processing behavior group about pea production technology. There was significant difference in information processing behavior between marginal and small farmers about pea cultivation technology.

Keywords: *Pea production, Processing, Behaviors, Hadoti, Rajasthan.*

Pea is a self pollinated crop and belongs to the family papilionaceae (Fabaceae). The fruit which is eaten is a 'pod' botanically. The plant is a harbaceous annual with vining habit. The leaves are pinnately compound with one to three pairs leaflets. The terminal leaflet and occasionally the upper lateral are modified as tendrils. The tap root grows up to 120 cm depending on soil type. Root nodules form on roots by bacteria, *Rhizobium leguminosarum.* Legume bacteria fix nitrogen from atmosphere in the soil.

Pea has been grown in India for several decades and is quite adapted to this part of the world. There are good number of local strains and exotic varieties available. Most of the exotic varieties were introduced in India and used as commercial varieties. Genetic diversity within the genus *pisum* has resulted in a myraid of uses for the crop. Dry peas are used for animal feed and for human consumption in soups and processed snacks.

Pea is the major vegetable crop of *rabi* season in Rajasthan. It occupies on area of 3729 hectares with the production of 5807 tonnes in Rajasthan (Vital Statistics, 2009-10). It is mainly cultivated in Jaipur, Nagaur, Kota, Tonk, Bundi, Alwar, and Chittor districts in the state. The Kota region of Rajasthan is leading in area and production of pea. It occupies on an area of 1176 hectares with the production of 1493 tonnes in Kota region (Vital Statistics, 2009-10). Kota region consists of Kota, Baran, Bundi, Jhalawar and Tonk districts. The soil and climatic condition of this region is most suitable for pea cultivation, while the productivity is far below as compared to recommended by the scientists. The low productivity of pea in Kota region of Rajasthan may be due to the poor information processing behaviour of the farmers. Looking to the above facts the present study was conducted with the following objectives:

1. To find out the information processing behaviours of the pea growers.
2. To see the significance difference about information processing behaviour of the pea growers.

Methodology

The present study was conducted in Kota region of Rajasthan. Kota region consist of five districts, out of which three districts namely Bundi, Kota and Tonk were selected purposively on the basis of maximum area under pea cultivation. Two tehsils from each identified districts were selected on the basis of maximum area under pea cultivation. Thus, in all six tehsils were taken for the present study. Total twenty villages were identified on the basis of proportionate sampling from the selected tehsils. To select the respondents, a comprehensive list of all pea growers was prepared for all villages. Thereafter, the farmers were categorized into two groups *i.e.* small and marginal on the basis of pea cultivation. The respondents selected randomly from each category of the farmers. It was planned to select 10 respondents *i.e.* five in each category from the each selected village. Thus, the total sample size of the study was 200 respondents. Data were collected through pre-structured interview schedule. Thereafter, data were analysed, tabulated and interpreted the results.

Results and Discussion

To get an overview of the pea growing farmers regarding processing behaviour for the received information, they were grouped into three strata as, low, medium and high on the basis of calculated mean and standard deviation of information processing score obtained by the respondents.

The data incorporated in Table 19.1 depict that out of 200 respondents, 45.50 per cent respondents had low information processing behaviour. This was followed by 31.50 per cent farmers having medium level of information processing behavior. Whereas, only 23.00 per cent respondents were found to be high information processing behavior group. Perusal of table further reveals that 51.00 and 40.00 per cent marginal and small farmers respectively were in the low information processing behaviour group. Likewise, 28.00 and 35.00 per cent marginal and small farmers had medium information processing behaviour about improved pea cultivation practices respectively. Whereas, 21.00 per cent marginal farmers and 25.00 per cent small farmers were found in high information processing behaviour group about pea production technology. Majority of farmers were found in low information processing behaviour group. It may be due to reason that majority of the farmers are illiterate and have small size of land holding and they are poor adopters of the innovation.

Table 19.1. Distribution of Respondents According to their Information Processing Behaviour about Pea Production Technology

n = 200

Sl. No.	*Degree of Information Processing Behaviour*	*Marginal Farmers*		*Small Farmers*		*Total*	
		Frequency	*Per cent*	*Frequency*	*Per cent*	*Frequency*	*Per cent*
1.	Low (< 21)	51	51.00	40	40.00	91	45.50
2.	Medium (21-32.5)	28	28.00	35	35.00	63	31.50
3.	High (> 32.5)	21	21.00	25	25.00	46	23.00
	Total	**100**	**100.00**	**100**	**100.00**	**200**	**100.00**

The findings are in agreement with those of Pramilla (1992) who reported that majority of the respondents possessed medium level of information processing behaviour. The extent of information processing behaviour of farmers was studied under three major modes *i.e.*, information evaluation methods, information storage methods and information transfer methods. The results of the same are presented in subsequent tables.

Information Evaluation Methods Used by the Pea Growers

The data incorporated in Table 19.2 indicate that the received information was weigh the merit of an innovation in the light of past experience by the marginal and small farmers to the extent of 93.67 and 97.73 MPS respectively and ranked first by both the categories of farmers. Likewise, the information received by the marginal and small pea growers were found to have judged on the basis of its economic feasibility which was accorded second rank to the extent of 92.73 and 96.23 MPS respectively.

Table 19.2. Information Evaluation Methods Used by the Pea Growers

n = 200

Sl. No.	*Information Evaluation Method*	*Marginal Farmers*		*Small Farmers*		*Total*	
		MPS	*Rank*	*MPS*	*Rank*	*MPS*	*Rank*
1.	Discussion with officials of State Department of Agriculture/ Agriculture University	25.87	8	27.17	8	26.52	8
2.	Acceptance of received information with modification	53.23	5	59.15	5	56.19	5
3.	Judgment on the basis of economic feasibility	92.73	2	96.23	2	94.48	2
4.	Acceptance of received information as such	43.21	6	48.67	6	45.94	6
5.	Discuss with family members, friends, fellow farmers, progressive farmers and neighbours	91.67	3	95.90	3	93.78	3
6.	Judgment in the light of climatic conditions	73.52	4	79.27	4	76.39	4
7.	Judgment based on technical feasibility	39.13	7	44.37	7	41.75	7
8.	Weigh the merit of an innovation in the light of past experience	93.67	1	97.73	1	95.70	1

MPS: Mean Per cent Score.

Further analysis of table shows that the received information was accepted after discussion with family members, friends, fellow farmers, progressive farmers and neighbours by marginal and small farmers to the extent of 91.67 and 95.90 MPS respectively and ranked third position by the pea growers. It was also found that judgment in the light of climatic conditions and acceptance of received information with modifications was accorded fourth and fifth rank respectively by both the categories of pea farmers. Analysis of table further shows that acceptance of information as such was placed on the sixth rank by marginal farmers and small farmers with 43.21 and 48.67 MPS respectively. The judgment based on technical feasibility was accorded seventh rank by both the categories of respondents in the order of information evaluation methods used by the respondents. A very small number of farmers discussed with officials of state Department of Agriculture/ Agriculture University and was ranked eighth by marginal and small farmers to the extent of 25.87 and 27.17 per cent respondents in the study area.

Thus, from the above discussion, it could be concluded that extent of information an evaluation method used by the marginal farmers was from 25.87 to 93.67 per cent, while small farmers was from 27.17 to 97.73 per cent with regard to pea cultivation technology.

Information Storage Methods Used by the Pea Growers

The data incorporated in Table 19.3 reveal that marginal and small pea growers stored the information received by them by way of memorizing the information with the extent of 85.93 and 92.21 MPS respectively and was ranked first by both the categories of farmers.

Table 19.3. Information Storage Methods Used by the Pea Growers

n = 200

Sl. No.	Information Storage Method	Marginal Farmers		Small Farmers		Total	
		MPS	Rank	MPS	Rank	MPS	Rank
1.	Conveying to family members and asking them to remember	72.60	2	76.97	2	74.78	2
2.	By maintaining classified notebooks/diary	16.37	3	23.40	3	19.88	3
3.	Preservation in the form of printed literature	9.45	4	19.85	4	14.65	4
4.	By maintaining subject matter file	8.14	5	17.03	5	12.58	5
5.	Memorizing the information	85.93	1	92.21	1	89.07	1

MPS: Mean Per cent Score.

This was followed by way of conveying to family members and asking them to remember which was placed at second position in the rank hierarchy of information storage methods. The extent of information storage by this method was 72.60 and 76.97 MPS among marginal and small farmers respectively. Further analysis of table also shows that received information was stored through maintaining classified note books/diary by the marginal farmers and small farmers with the extent of 16.37 and 23.40 per cent respectively and it was accorded third rank by marginal and small farmers. The storage of information by preserving in the form of printed literature was assigned fourth rank by marginal and small farmers with the extent of 9.45 and 19.85 MPS respectively. The storage of information by maintaining subject matter file was very less used and accorded last rank by the respondents with the extent of 8.14 and 17.03 per cent among marginal and small farmers respectively. It was observed during the period of data collection that majority of the farmers did not find time after the farm activities and their house hold work. This might be the reason that majority of the respondents stored the received information by memorizing and conveying to family members.

The findings are in accordance with those of Pramilla (1992) who reported that majority of the respondents stored the received information by conveying to family members and by memorizing.

Information Transfer Methods Used by the Pea Growers

The data presented in Table 19.4 reveal that marginal and small pea growers transferred the information to relative with the extent of 65.09 and 73.50 per cent

MPS respectively and ranked first by marginal and small farmers. Similarly, the marginal and small farmers transferred the information to their friends, fellow farmers, progressive farmers and neighboures with extent of 61.19 and 69.13 MPS respectively and ranked second by both the categories of farmers. It was observed during the period of data collection that majority of the pea growers had regular and face to face contact with their friends, fellow farmers and neighboures in their locality. This might be the reason that majority of them transferred the received information to their friends, fellow farmers and neighbours.

Table 19.4. Information Transfer Methods Used by the Pea Growers

n=200

Sl. No.	*Information Transfer Method*	*Marginal Farmers*		*Small Farmers*		*Total*	
		MPS	*Rank*	*MPS*	*Rank*	*MPS*	*Rank*
1.	Those who come to seek	57.47	3	65.16	3	61.31	3
2.	To friends, fellow farmers, progressive farmers and neighbours	61.19	2	69.13	2	65.16	2
3.	To relatives	65.09	1	73.50	1	69.29	1
4.	To those who cultivate their land on lease	53.40	4	61.96	4	57.68	4
5.	Speaking in local meetings	21.83	5	27.57	5	24.70	5
6.	By conducting demonstrations to show the practical aspect of received information	9.93	6	14.43	7	12.18	6
7.	Lending printed literature to others	8.12	7	15.01	6	11.56	

MPS = Mean Per cent Score

Further, analysis of Table 19.4 also reveals that marginal and small farmers provided the received information to those who come to seek with the extent of 57.47 and 65.16 per cent respectively. Likewise, the method of giving information to those who cultivated their land on lease was assigned fourth rank by the marginal farmers and small farmers with the extent of 53.40 and 61.67 per cent respectively. The extent of information transferred through speaking in local meeting was 21.83 and 27.57 per cent by the marginal and small farmers respectively.

It was also found that method of conducting demonstrations to show the practical aspect of received information was accorded sixth rank by marginal farmers and seventh by small farmers. Whereas, the small proportion of marginal and small farmers had the habit of lending printed material to others. This may be due to the reason that majority of the farmers were not educated highly in the study area so that they were not read the information of pea cultivation technology through printed literature. From the above discussion, it could be concluded that the extent of information transfer methods used by the marginal farmers was from 8.12 to 65.09 per cent and among small farmers it was noted from 14.43 to 73.50 per cent.

These findings are in agreement with those of Pramilla (1992) and Ramasubramanian and Manoharan (2003) who reported that majority of the respondents transferred the received information to their friends, fellow farmers and neighbourers.

Comparison of Information Processing Behaviour of the Pea Growers

The comparison of information processing behaviour of the pea growers was made under following heads:

Comparison of Information Processing Behaviour between Marginal and Small Farmers

To study the difference of information processing behaviour between marginal and small respondents about pea cultivation technology, 'Z' test was applied. The results are presented in Table 19.5.

Hypotheses

NH_{01}: There is no significant difference between marginal and small farmers about information processing behaviour of pea cultivation technology.

RH_1: There is significant difference between marginal and small farmers about information processing behaviour of pea cultivation technology.

Table 19.5. Significance of difference about Information Processing Behaviour of Pea Growers

Sl.No.	*Category of Respondents*	*Mean*	*S.D.*	*'Z' value*
1.	Marginal farmers	18.58	5.2	2.07*
2.	Small farmers	20.21	5.9	

* Significant at 5 per cent level.

Table 19.5 shows that the calculated 'Z' value was found to be greater than its tabulated value at 5 per cent level of significance. Thus, the null hypothesis (NH_1) was rejected and alternate hypothesis (RH_1) was accepted. It reveals that there was significant difference in information processing behaviour between marginal and small farmers about pea cultivation technology. The mean value further indicates that small farmers had higher information processing behaviour than marginal farmers about pea cultivation technology. This difference in the level of information processing behaviour of pea growers might be due to the reason that small respondents had more socio-economic status, more extension contact, high participation in training programmes and more mass media exposure as compared to marginal pea growers.

These findings are in line with those of Kaidan and Kumar (2002) and Vashishtha (2007) who reported that there was a significant difference between different categories of respondents with respect to their information processing behaviour.

Comparison of Information Processing Behaviour between Selected Districts

In relation to information processing behaviour of pea cultivation, it was also felt necessary to study the difference between farmers of selected districts. To find out the variation in information processing behaviour of the farmers of selected districts analysis of variance test ('F' test) was applied. The results are presented in Table 6.

Hypotheses

NH_2: There is no significant difference in information processing behaviour between the pea growers of selected districts.

RH_2: There is significant difference in information processing behaviour between the pea growers of selected districts.

Table 19.6. Comparison of Information Processing Behaviour between the Pea Growers of Selected Districts

n=200

Sl.No.	*Source of Variation*	*d.f.*	*S.S.*	*M.S.S.*	*'F' cal*
1.	Between the districts	2	2784.22	1392.11	3.41*
2.	Error	197	80418.78	408.22	
	Total	**199**	**83203.00**		

* Significant at 5 per cent level

Table 19.66 reveals that the calculated 'F' value 3.41 which is higher than the tabulated value at 5 per cent level of significance and 2 degree of freedom. Thus, the null hypothesis (NH_{02}) entitled "there is no significant difference in the pea growers of selected districts about information processing behaviour was rejected and alternative hypothesis (RH_2) was accepted. Thus, it is concluded that there was significant difference in information processing behaviour of the pea growers of selected districts towards pea production technology. By comparing the mean value with C.D. value, there is significant difference between Tonk and Kota districts and also between Bundi and Tonk districts but non-significant difference was observed between Kota and Tonk districts with regard to information processing behaviour. The respondents of Bundi district possessed more information processing behaviour than respondents of Kota and Tonk districts.

Mean Value Table

Sl.No.	*Name of Dstricts*	*Mean Value*	*C.D. Value*
1.	Bundi	20.25	1.116
2.	Kota	19.42	
3.	Tonk	16.70	

Conclusions

Findings revealed that out of 200 respondents, 45.50 per cent respondents had low information processing behaviour and 31.50 per cent farmers having medium level of information processing behaviour. Whereas, only 23.00 per cent respondents were observed in high information processing behaviour group. Results further revealed that 51.00 and 40.00 per cent marginal and small farmers respectively were in the low information processing behaviour group. Likewise, 28.00 and 35.00 per cent marginal and small farmers had medium information processing behaviour about improved pea cultivation practices respectively. Whereas, 21.00 per cent marginal farmers and 25.00 per cent small farmers were found in high information processing behaviour group about pea production technology. From the above results it can be concluded that 45.50 per cent respondents had low information processing behaviour and 31.50 per cent farmers having medium level of information processing behaviour. Whereas, only 23.00 per cent respondents were observed in high information processing behaviour group. There was significant difference in information processing behaviour between marginal and small farmers about pea cultivation technology.

REFERENCES

Chaturvedi, D., Dangi, K.L. and Sharma, F.L. (2009). Technological knowledge of cauliflower growers in Udaipur district of Rajasthan. In: *Proceedings of National Seminar on Information Technology Application in Agriculture for Livelihood Security of Farmers,* organized by Rajasthan Society of Extension Education and DEE, Udaipur from Nov. 10-12, pp. 43.

Chauhan, A.S. (1994). Adoption behaviour of tribal farmers towards pigeon-pea technology. *Maharashtra Journal of Extension Education* **13**: 43-46.

Chauhan, N.B. (1994). A study of peasentry modernization in Dungarpur district of Rajasthan. Ph.D. thesis submitted to Rajasthan Agricultural University, Bikaner.

Choudhary, B.N. (1991). Training strategy for promoting rainfed agriculture technologies in India. *Paper published in compendium of national seminar of extension strategies for rainfed agriculture* organized by Indian Society of Extension Education, ICAR, New Delhi: 78.

Desai, C.P., Pandey, D.N., Patel, M.R. and Patel, A.A. (1996). Farmers' satisfaction with adoption of cumin cultivation. *Gujarat Agricultural University Research Journal* **21**: 75-76.

Deshmukh, S.K., Shinde, P.S. and Bhople, R.S. (1997). Adoption of summer groundnut production technology by the growers. *Maharashtra Journal of Extension Education* **16**: 326-329.

Dhinga, A and Hasija, R.C. (2003). A study of farmer's awareness regarding agricultural production in Punjab. *Agricultural Extension Review* **15**: 16-21

Kaidan, K.S. and Kumar, R. (2002). Information processing pattern of dairy farmers of Kangra Vally. *Indian Journal of Extension Education* **38**: 65-67.

Pramilla, K. (1992). An expost facto study on communication channels utilization behaviour of farm women. M.Sc. Thesis submitted to Tamil Nadu Agricultural University, Coimbatore.

Ramasubramaniam, M. and Manoharan, M. (2003). Communication behaviour of mango growers. *Agriculture Extension Review* **15**: 16-20.

Vashishtha, U. (2007). Farmers' response towards pigeon-pea (*Cajanus cajan* L.) cultivation in Tribal area of Udaipur district of Rajasthan. M.Sc. (Ag.) Thesis submitted to the Maharana Pratap University of Agriculture and Technology, Udaipur, Rajasthan.

Chapter 20

Adoption Behaviour of Farmers in Pea Production: A Case in Hadoti Region of Rajasthan

N.R. Meena, Supriya and Ekta Pandey

Assistant Professor, College of Agriculture, ANDUAT, Kumarganj, Ayodhya, Uttar Pradesh

The present study was conducted on the 200 farmers in Hadoti region ((Bundi, Kota and Tonk districts) of Rajasthan to identify adoption behavior of new technology of pea production. The findings revealed that 105 (52.50 per cent) of total pea growers were found to be from medium adoption level group, whereas, 43 (21.50 per cent) respondents were reported from the group of low adoption level and 52 (26.00 per cent) respondents were in the high adoption level. The study further revealed that 23.00 per cent marginal farmers were in the high adoption level and 28.00 per cent marginal farmers were in the low adoption level, while 49.00 per cent marginal pea growers were found in the medium adoption category. In case of small farmers' category it was observed that 15.00, 56.00 and 29.00 per cent respondents had low, medium and high level of adoption respectively about pea production practices in the study area. It was noted that the extent of adoption in marginal farmers was from 23.70 to 73.20 per cent, while in case of small farmers the extent of adoption was from 30.37 to 74.12 per cent in all improved pea cultivation practices. Hence, it is concluded that there was significant difference between marginal and small farmers in adoption of improved pea production technology.

***Keywords**: Adoption, Pea, Hadoti, Farmers, Medium, Large, Rajasthan.*

Pea is excellent food for consumption, taken either as a vegetable or in soup. Large proportion of pea is processed (canned, frozen or dehydrated) for consumption in the off season. Pea straw is a nutritional fodder. Pea is a rich source of protein

(25 per cent), amino acids and sugars (12 per cent) alongwith carbohydrates and vitamins. Oil from ripened seed has antisex harmonic effects, produces sterility and antagonizes effect of male harmone. Pea seeds contain trypsin and chymotrypsin which could be used for contraceptive, ecbolic, fungistatic and spermicide. Leaves are used as a pot herb in India, Burma and parts of Africa.

Pea is a cool season crop best suited to production in the temperate regions. In the warmer areas of the tropics pea is restricted to production in the cooler highlands. Temperature between 7° and 24°C are suitable for plant growth but optimum yield are achieved between 13° and 21°C. It is relatively tolerant to frost when compared with solanaceous and cucurbitaceous crops. Blossoms and pods are susceptible to frost, whereas leaves and stems are relatively tolerant. Seed germinates better at the soil temperature of 10-18.3°C. The pea crop is produced primarily under rainfed conditions but can also be irrigated. Pea has been grown in India for several decades and is quite adapted to this part of the world. There are good number of local strains and exotic varieties available. Most of the exotic varieties were introduced in India and used as commercial varieties. Genetic diversity within the genus *pisum* has resulted in a myraid of uses for the crop. Pea can be grown on all types of soils but the early crop is expected in light soils and higher yields in loose, friable and heavy soils. Pea is grown as a rabi crop and sown from the beginning of October to middle of November in the Kota region of Rajasthan.

The Kota region of Rajasthan is leading in area and production of pea. It occupies on an area of 1176 hectares with the production of 1493 tonnes in Kota region (Vital Statistics, 2009-10). Kota region consists of Kota, Baran, Bundi, Jhalawar and Tonk districts. The soil and climatic condition of this region is most suitable for pea cultivation. Looking to the above facts, the present study was conducted in Kota region of Rajasthan with specific objectives:

1. To find out the level of adoption of improved pea cultivation practices by the farmers.
2. To compare the level of adoption of improved pea cultivation technology between the small and marginal farmers.

Materials and Methods

The present study was conducted in Kota region of Rajasthan. Kota region consist of five districts, out of which three districts namely Bundi, Kota and Tonk were selected purposively on the basis of maximum area under pea cultivation. Two tehsils from each identified districts were selected on the basis of maximum area under pea cultivation. Thus, in all six tehsils were taken for the present study. Total twenty villages were identified on the basis of proportionate sampling from the selected tehsils. To select the respondents, a comprehensive list of all pea growers was prepared for all villages. Thereafter, the farmers were categorized into two groups *i.e.* small and marginal on the basis of pea cultivation. The respondents selected randomly from each category of the farmers. It was planned to select 10 respondents *i.e.* five in each category from the each selected village. Thus, the total sample size of the study was 200 respondents. Data were collected through pre-structured interview schedule. Thereafter, data were analysed, tabulated and interpretated in the light of the objective.

Results and Discussion

To get an overview of the respondents with respect to their level of adoption, they were grouped into three strata *viz.* low, medium and high adoption group. This stratification was based on the calculated mean value and standard deviation of the adoption score obtained by the respondents. The results are presented Table 20.1.

Table 20.1. Distribution of Pea Growers on the Basis of their Level of Adoption of Improved Pea Production Technology

n = 200

Sl. No.	*Level of Adoption*	*Marginal Farmers*		*Small Farmers*		*Total*	
		Frequency	*Per cent*	*Frequency*	*Per cent*	*Frequency*	*Per cent*
1.	Low (< 52)	28	28.00	15	15.00	43	21.50
2.	Medium (52-77)	49	49.00	56	56.00	105	52.50
3.	High (> 77)	23	23.00	29	29.00	52	26.00
	Total	**100**	**100.00**	**100**	**100.00**	**200**	**100.00**

The data in Table 20.1 reveal that 105 (52.50 per cent) of total pea growers were found to be from medium adoption level group, whereas, 43 (21.50 per cent) respondents were reported from the group of low adoption level and 52 (26.00 per cent) respondents were in the high adoption level. An analysis of small and marginal respondents regarding their level of adoption about improved pea production technology revealed that 23.00 per cent marginal farmers were in the high adoption level and 28.00 per cent marginal farmers were in the low adoption level, while 49.00 per cent marginal pea growers were found in the medium adoption category. In case of small farmers' category, it was observed that 15.00, 56.00 and 29.00 per cent respondents had low, medium and high level of adoption respectively about pea production practices in the study area.

It can be concluded that small farmers possessed more adoption level than marginal farmers about improved practices of pea cultivation. It may be due to the reason that knowledge level of small farmers was comparatively higher than that of marginal farmers that might have contributed to slightly higher level of adoption. Another reason behind such findings may be that the small farmers were slightly more progressive in nature than marginal farmers. Thus, adoption is dependent on knowledge and social participation might have also played vital role in higher adoption of small farmers.

Similar findings have been reported by Vashishta (2007) who found that 56.25 per cent of the total respondents adopted the pigeon pea production technology to a medium level, whereas, 26.88 and 16.87 per cent of the total respondents adopted pigeon pea production technology at low and high level respectively.

To find out the aspect wise extent of adoption among small and marginal pea growers about improved pea cultivation practices, mean per cent score of each major practice was calculated and presented in Table 20.2.

Table 20.2. Extent of Adoption of Improved Pea Production Technology among the Farmers

n = 200

Sl. No.	Improved Practices	Marginal Farmers		Small Farmers		Total	
		MPS	Rank	MPS	Rank	MPS	Rank
1.	Use of high yielding varieties	51.50	7	58.54	6	55.02	7
2.	Soil and field preparation	65.60	2	71.15	2	68.38	2
3.	Soil treatment	23.70	12	30.37	12	27.03	12
4.	Seed treatment	41.70	9	45.60	9	43.65	9
5.	Time of sowing	73.20	1	74.12	1	73.66	1
6.	Seed rate and spacing	60.54	4	62.50	4	61.52	4
7.	Fertilizer application	56.23	5	59.70	5	57.97	5
8.	Water management	53.70	6	57.95	7	55.82	6
9.	Weed management	61.67	3	65.20	3	63.44	3
10.	Staking	40.62	10	45.33	10	42.98	10
11.	Plant protection measures	40.10	11	41.33	11	40.72	11
12.	Harvesting, post-harvest techniques and marketing	48.67	8	51.80	8	50.23	8

MPS: Mean Per cent Score.

Data accorded in Table 20.2 indicate that highest adoption was observed in the practice of time of sowing among the marginal and small farmers with the extent of 73.20 and 74.12 per cent respectively. It was noticed that majority of the farmers were following the recommended time of sowing *i.e.* last week of October to first week of November. The study of table further reveals that the adoption regarding soil and field preparation was recorded 65.60 and 71.15 per cent among marginal and small farmers respectively and it was ranked second by both the categories of respondents. It means that majority of pea growers were following all important operations for soil and field preparation for successful cultivation of pea crop.

The extent of adoption about weed management was 61.97 and 65.20 per cent among marginal and small pea growers respectively. The adoption level of small farmers was higher than marginal farmers in weed management and intercultural operations in pea crop. The adoption of seed rate and spacing is almost equal among marginal and small farmers and ranked fourth in ranking hierarchy of all improved pea cultivation practices.

Analysis of table further reveals that adoption regarding fertilizer application among marginal and small farmers was 56.23 and 59.70 per cent respectively. Majority of the farmers of the study area were not applying various fertilizers in pea cultivation as per the recommendation of scientists. The extent of adoption about water management was found to be 53.30 and 57.95 MPS among marginal and small farmers respectively and it was ranked sixth by marginal farmers and seventh by small farmers. It was observed during analysis of data that recommended irrigation management aspects were using by majority of marginal and small farmers because ample of irrigation water is available in Kota region of Rajasthan.

In case of the extent of adoption regarding the high yielding varieties of pea was 51.50 and 58.54 per cent among marginal and small pea growers respectively. Analysis of data clearly shows that majority of the respondents were adopting the high yielding varieties namely Rachna, T-163, Pusa Ratna, which are more suitable and providing higher yield. It was further noted that extent of adoption in the case of harvesting, post-harvest techniques and marketing was 48.67 and 51.80 per cent among marginal and small farmers respectively. Majority of the farmers were not following the cold storage practices of pea because such facilities are not common in the study area. The adoption about seed treatment was noticed 41.70 and 45.60 per cent among marginal and small farmers respectively and it was ranked ninth by both the categories of pea growers.

Further analysis of table reveals that extent of adoption of staking technique in pea cultivation was 40.62 and 45.33 per cent among marginal and small farmers respectively and this aspect was placed on tenth position by both the categories of respondents. The extent of adoption of plant protection measures was observed to be 40.10 and 41.33 per cent among marginal and small respondents respectively. The adoption level of pea growers was not encouraging about this aspect. This may be due to the fact that chemicals used for controlling insect-pests and diseases are costly affairs. The extent of adoption about soil treatment aspect was prioritized at the last by the pea growers.

From the above discussion, it can be concluded that the extent of adoption in marginal farmers was from 23.70 to 73.20 per cent, while in case of small farmers the extent of adoption was from 30.37 to 74.12 per cent in all improved pea cultivation practices. Further it was found that small farmers had more extent of adoption than marginal farmers but there still exists a gap in the adoption of improved pea cultivation practices.

Comparison of Adoption of the Pea Growers between Different Components about Improved Pea Cultivation Technology

The comparison in the adoption of improved pea cultivation technology by the farmers was made under following heads:

Comparison of Adoption between Marginal and Small Farmers

To find out the variation or similarity in the adoption of improved pea cultivation technology between the marginal and small farmers, the 'Z' test was applied. The results of the same have been given in Table 20.3.

Hypotheses

NH_{01}: There is no significant difference between marginal and small farmers in adoption of improved pea production technology.

RH_1: There is significant difference between marginal and small farmers in adoption of improved pea production technology.

Table 20.3. Comparison of the Adoption between Marginal and Small Farmers Regarding Improved Pea Cultivation Technology

n = 200

Sl.No.	*Category of Respondents*	*Mean*	*S.D.*	*'Z' value*
1.	Marginal farmers	43.73	7.4	2.87**
2.	Small farmers	47.01	8.7	

** Significant at 1 per cent level.

Table 20.3 shows that the calculated value of 'Z' (2.87) is greater than the tabulated value at 1 per cent level of significance. This calls for the acceptance of the research hypothesis (RH_1) *i.e.*, there is significant difference between marginal and small farmers in adoption of improved pea production technology. Hence, it is concluded that there was significant difference between marginal and small farmers in adoption of improved pea production technology. The mean value further indicates that the small farmers had higher adoption mean than marginal farmers. This difference in the level of adoption between small and marginal farmers might be due to the reason that small respondents had more cosmopolitan outlook, economic motivation, extension contact, size of land holding and income level than marginal farmers.

Comparison of Adoption between Selected Districts

Analysis of variance test ('F' test) was applied to find out the significance of variation in adoption of the farmers of selected districts. The results of ANOVA computed for this purpose are presented in Table 20.4.

Hypotheses

NH_{02}: There is no significant variation in adoption of improved pea cultivation practices by the farmers of selected districts.

RH_2: There is significant variation in adoption of improved pea cultivation practices by the farmers of selected districts.

Table 20.4. Comparison of Aadoption Score between the respondents of selected districts

n = 200

Sl.No.	*Source of variation*	*d.f.*	*S.S.*	*M.S.S.*	*'F' cal*
1.	Between the selected districts	2	7890.28	3945.40	11.63**
2.	Error	197	66846.72	339.32	
	Total	199	82203		

** Significant at 1 per cent level of significance.

Mean value table

Sl.No.	Name of Districts	Mean Value	C.D. Value
1.	Bundi	45.58	1.108
2.	Kota	43.20	—
3.	Tonk	41.65	—

Data presented in Table 20.4 show that the calculated 'F' value (11.63) is higher than the tabulated 'F' value at 1 per cent level of significance and 2 degree of freedom. So the result is statistically significant. Thus, Null hypothesis (NH_{02}) was rejected and research hypothesis (RH_2) was accepted. It means that there was a significant variation between the respondents of selected districts with regard to adoption of improved pea cultivation practices. By comparing the mean value with C.D. value, there was significant difference among Bundi, Kota and Tonk with respect to adoption of improved pea production technology.

Conclusions

The findings revealed that 105 (52.50 per cent) of total pea growers were found to be from medium adoption level group, whereas, 43 (21.50 per cent) respondents were reported from the group of low adoption level and 52 (26.00 per cent) respondents were in the high adoption level. The study further revealed that 23.00 per cent marginal farmers were in the high adoption level and 28.00 per cent marginal farmers were in the low adoption level, while 49.00 per cent marginal pea growers were found in the medium adoption category. In case of small farmers' category it was observed that 15.00, 56.00 and 29.00 per cent respondents had low, medium and high level of adoption respectively about pea production practices in the study area. It was noted that the extent of adoption in marginal farmers was from 23.70 to 73.20 per cent, while in case of small farmers the extent of adoption was from 30.37 to 74.12 per cent in all improved pea cultivation practices. Further it was found that small farmers had more extent of adoption than marginal farmers but there still exists a gap in the adoption of improved pea cultivation practices. The findings indicated that there was significant difference between marginal and small farmers in adoption of improved pea production technology. Hence, it is concluded that there was significant difference between marginal and small farmers in adoption of improved pea production technology.

It was concluded that the extent of adoption in marginal farmers was from 23.70 to 73.20 per cent, while in case of small farmers the extent of adoption was from 30.37 to 74.12 per cent in all improved pea cultivation practices. Further it was found that small farmers had more extent of adoption than marginal farmers but there still exists a gap in the adoption of improved pea cultivation practices.

REFERENCES

Badodia, S.K., Shrivastava, K.K. and Lakhera, M.L. (2002). Technological gap in chickpea cultivation technology. *Agricultural Extension Review* **14**: 25-28.

Bareth, L.S. (1991). Technological constraints in adoption of improved pulse production technologies in agro-climatic zone II A and B of Rajasthan. Ph.D. thesis submitted to Rajasthan Agricultural University, Bikaner, Rajasthan.

Chundawat, M.S. (1997). Knowledge and adoption of improved practices of cumin production technology in the Jalore district of Rajasthan. M.Sc. (Ag.) thesis submitted to Rajasthan Agricultural University, Bikaner.

Kubde, V.R., Tekale, V.S. and Bhople, R.S. (1999). Knowledge and adoption of soyabean production technology of farmers. *Maharashtra Journal of Extension Education* **18**: 185-188.

Kumar, M. (2004). Status and future strategy of gram (*Cicer arietinum* L.) cultivation in Jhunjhunu District of Rajasthan. M.Sc. thesis submitted to Maharana Pratap University of Agriculture and Technology, Udaipur, Rajasthan.

Kumar, M. (2008). A critical analysis of problems and prospects of groundnut cultivation in southern plain and aravalis hills (Zone IV a) of Rajasthan. Ph.D. thesis submitted to Maharana Pratap University of Agriculture and Technology, Udaipur.

Kumar, S. (1997). Adoption of turmeric (*Curcuma longa* L.) production technology by the farmers of Jhadol panchayat Samiti in Udaipur district of Rajasthan. M.Sc. (Ag.) Thesis submitted to Rajasthan Agricultural University, Bikaner.

Kumar, V.P. and Karippai, Jakhar, P.R. (2009). Adoption of improved cultivation practices of moongbean [*Vigna radiata* (L.) Wilesek] by small and marginal farmers of Nagaur District of Rajasthan. M.Sc. (Ag.) thesis submitted to Rajasthan Agricultural University, Bikaner.

Patel, B.D. (2005). A study on adoption of recommended chilli technology in Vadodra district of Gujarat state. M.Sc. (Ag.) thesis submitted to Gujarat Agricultural University, Anand.

Chapter 21

Assessment of Soil Quality Under Rice-Wheat Farming System of Soil of Thakurdwara Tahsil of Moradabad District, Uttar Pradesh

Ravindra Kumar, Manoj Singh, N.C. Tripathi, Devendra, Pal Singh and A.K. Mishra

Krishi Vigyan Kendra, Moradabad-II, KVK, Rampur, KVK, Shahjahapur, KVK, Sambhal, KVK, Amroha, SVPUA&T, Meerut, Uttar Pradesh

Macro and micro nutrients are important soil elements that control its fertility. Soil fertility is one of the important factors controlling yields of the crops. Soil characterization in relation to evaluation of fertility status of soil of an area or region is an important aspect in context of sustainable agriculture production. Because of imbalanced and inadequate fertilizer use coupled with low efficiency of other inputs, the response efficiency of chemical fertilizer nutrients has declined tremendously under intensive agriculture in recent year. In the present investigation, an attempt has been made to examine the chemical properties of soil in rice - wheat farming system. The study area covers Thakurdwara Thasil of Moradabad district of Uttar Pradesh. Soil samples of 0-15 cm depth were collected from 326 sites covering 21 gram panchayats. Collected soil samples were air dried in shade, crushed gently with a wooden roller and pass through 2.0 mm sieve to obtain a uniform representative sample. The processed soil samples were analyzed by standard methods.The pH varied from 5.2 to 9.2, organic carbon content varied from 3.9 to 6.9 g kg^{-1} soil. The available N content was varied from 156.96 to 259.32 kg ha^{-1} with an average value of 224.32 kg ha^{-1} The available phosphorous content varied from 21.79 to 56.53 P_2O_5 kg ha^{-1} with a mean value of 37.18 P_2O_5 kg ha^{-1}.Status of available potassium in the ranged from 158.20 to 283.25 K_2O Kg ha^{-1} with an average value of 211.92 K_2O kg ha^{-1}. Cu in the surface soil was found to sufficient and varied from 0.258 to 1.708 mg kg^{-1} The iron content varied *from 3.214 to 16.852, Mn from 1.701 to 8.351 mg kg^{-1}. The available Zn in surface (0-15 cm) in soil ranged from 0.425 to 1.708 mg kg^{-1} soil in rice-wheat farming system. Nutrient status regarding to the available macro and micro nutrient in surface soil indicate that soils are low in available N and medium in available P and K and in general marginal in available Cu,Fe, Mn and Zn. Normal to slightly alkaline in reaction, low to medium in organic carbon content.*

Keywords: *Soil fertility, Macro- and micro-nutrients, Availability.*

According to the International Food Policy Research Institute, between 1993 and 2020 A.D. the global demand for cereals is expected to increase by 41 per cent. It has been projected that annual rice production must increase from 556 million tons in 2000 A.D. to 758 million tons by 2020 A.D., a 36 per cent increase. Rice-wheat cropping system (RWCS) is a long-established grain production system in China. The wheat yield following rice was only 0.7 to 1.0 tons ha^{-1} until the 1940s and it increased progressively after the 1950s as a result of improved varieties, better agronomic management, and pest control. RWCS in the Indian subcontinent is quite new and started only in the late 1960s with the introduction of dwarf wheat from CIMMYT, Mexico, which required a lower temperature for good germination than that required for traditional tall Indian wheat. There could be many more variants involving vegetables and other short duration crops. Most rice in RWCS is transplanted and rice varieties grown are of 90–140 days duration (seed to seed) of which 25–45 days may be spent in nursery. The estimates of area under RWCS in the world vary considerably. In RWCS there is very little turn-around time between rice harvest and wheat sowing. Depending on the time of harvest of the rice crop, conventional tillage requires pre-sowing irrigation on well-drained soils or draining or drying of soil in lowlands followed by one or two disking, two harrowing, and leveling. Wheat in the RWCS belt in Indo-Gangetic Plains (IGP) is an irrigated crop. Considerable research has been conducted in India on the irrigation of wheat. Currently, there is a growing concern in sustainability of RWCS as the growth rates of rice and wheat yields are either stagnant or declining. Studies on socioeconomic and policy factors on the productivity of RWCS will be effective for measuring outcome of RWCS.

With increasing demographic pressure coupled with scarcity of soil and water resources, sustainable agriculture is not synonymous with "low-input" or organic agriculture. In some cases, low-input system may be acceptable for a short time, but in others like major food grain crops it may not be acceptable at all. As there is no alternative to agricultural intensification in our country, we must ensure using soil resources as per their capability and adopting the practices that improve soil quality and maintain a favorable soil condition for plant growth and environmental health (Mishra, 2005).

Use of high yielding varieties, intensive cropping, increase use of high analysis fertilizers and restricted use of organic sources of Nutrients has resulted in the deficiency of macro and micro nutrients in general particularly in the irrigated lands.(Ratan and Sharma, 2004). Nutrient removed by crop depends on cultivar, soil moisture status, management levels and residue management. Macro and micro nutrients are important soil elements that control its fertility. Soil fertility is one of the important factors controlling yields of the crops. Soil characterization in relation to evaluation of fertility status of soil of an area or region is an important aspect in context of sustainable agriculture production. Because of imbalanced and inadequate fertilizer use coupled with low efficiency of other inputs, the response efficiency of chemical fertilizer nutrients has declined tremendously under intensive agriculture in recent year. Recent diagnostic survey indicate that in many intensively cultivated area farmers have resorted to use greater than recommended doses of fertilizer, especially N Fertilizer, to maintained the crop productivity at levels attained previously with relatively small fertilization rates. This is an indication of decline in factor productivity. Low fertility of India soil is the major constant in achieving high productivity goals. In both agriculturally advanced irrigated ecosystems, nutrient replenishment through fertilizers and manures remain far below crop removal, thus causing the mining of native nutrient reserves over year. Wide spread deficiencies of macro and micro nutrients have emerged, and significant crop response to application of these nutrients are reported. The deficiencies are so intense and severe that visual symptoms are very often observed in major crops (Kumar et.al. 2013**)**. The results of numerous field experiments in different parts of India have, therefore indicated *"Fertilizer-induced unsustainability of crop productivity"* (Yadav 2003). Variation in nutrient supply is a natural phenomenon and some of them may be sufficient where other deficient. The stagnation in crop productivity cannot be boosted without balanced and optimal dose of inorganic fertilizers use of organic such as farm yard manure, compost, green manure, crop residue incorporation use of industrial waste biofertilizer, N fixers both symbiotic and associate and p solubilizers. Variations in nutrient supply are a natural phenomena and some where may be sufficient while some where deficient. Within a soil, variability may exist depending upon the hydrological properties of the soil and cropping system therefore 21 locations will required different management practices to sustained crop productivity and for this full information about the nutrient status is important. Therefore to have sound information about the nutrient status of these soils this study was under taken.

Material and Method

Moradabad, city, northern Uttar Pradesh state, northern India. It is situated on a ridge along the Ramganga River (a tributary of the Ganges (Ganga) River), about 24 km west-northwest of Rampur. The city, located at a major road and rail junction, is a trade centre for agricultural products. Industries include cotton milling and weaving, metalworking, electroplating, and printing. Moradabad's surrounding region consists of a level plain bounded by the Ganges River on the west and drained by the Ramganga. Grains, cotton, and sugarcane are grown.

The study area covers Thakurdwara Thasil of Moradabad district of Uttar Pradesh. Soil samples of 0-15 cm depth were collected from 326 sites covering 21 gram panchayats. Collected soil samples were air dried in shade, crushed gently with a wooden roller and pass through 2.0 mm sieve to obtain a uniform representative sample. Samples were properly labeled with the aluminum tag and stored in polythene bags for analysis.The processed soil samples were analyzed by standard methods for pH and electrical conductivity (1:2 soil water suspensions), organic carbon (Walkley and Black, 1934), available nitrogen (Subbiah and Asija, 1956), available phosphorus (Olsen *et al.*, 1954), available potassium (Jackson, 1973) and available micronutrients (Fe, Mn, Zn and Cu) in soil samples with extracted diethylene triamine penta acetic acid (DTPA) solution (0.005M) DTPA+0.01M $CaCl_2$ +0.1M triethanolamine, pH 7.3 as outlined by Lindsay and Norvell (1978).

Results and Discussion

The soil samples collected from different locations of Thakurdwara, Dilari, Bhagatpur Tanda and Chaijlait of Moradabad district at surface and subsurface soil where sugarcane - wheat farming system was followed by farmers usually apply 140-150 kg N/ha along with 60-75 kg P/ha. Zinc application in rice-wheat farming system done by 45 per cent of farmers while green manuring practiced by 10-12 per cent farmers and biofertilizers use was not prevalent. It was noted that 85 per cent farmers reported increased use of fertilizers to harvest same quantity of yield at different locations of Thakurdwara Tahsil of Moradabad districts of Uttar Pradesh.

Chemical Properties

It was observed that soil pH varied from 5.2 to 9.2 with an average of 7.3 according to classification of soil reaction suggested by Brady (1985), 25 samples were normal (7.2 to 7.3), 47 samples were mildly alkaline (pH 7.4 to 7.8), 70 samples were moderately alkaline (pH 7.9 to 8.2). The minimum value of pH 5.2 was observed in Padla and Maximum value of pH 9.2 was observed in Lalabala, and Ratupura. The relatively high pH of soils might be due to the presence of high degree of base saturation. The electrical conductivity of the soil varied from 0.120to 0.989 dSm^{-1}.

Organic Matter Content

Organic carbon content of the soil in rice-mentha+wheat varied from 3.9 to 6.9 g Kg^{-1} soil.The organic carbon content was low (<0.50 per cent) in 26 per cent, medium (0.5 to 0.75 per cent) in 74 per cent soil samples. High temperature and more tillage practice in the soil increases the rate of oxidation of organic matter resulting reduction of organic carbon content. Agarwal *et al.* (1990) reported that organic carbon content of some soil Rajasthan ranged from 0.142 to 0.40 per cent.

Available Nitrogen Content

Soil fertility exhibits the status of different soils regard to the amount and availability of nutrients essentials for plant growth. The available N content in rice- wheat varied from 156.96 to 259.32 kg ha^{-1} with an average value of 224.32 kg ha^{-1} (Table 21.1). On the basis of rating suggested by Subbiah and Asija (1956), all samples were low (<250 N kg ha^{-1}) IN available nitrogen. A significant positive correlation (r =0.933) was found between organic carbon and available nitrogen. This relationship was found because most of the soil nitrogen is in organic form. Similar results were also reported by Verma *et al.* (1980).

Table 21.1. Salient Soil Properties (Weighted mean) under Rice-wheat Farming System

Sl. No.	*Name of Village*	*No of Samples Collected*	*pH*	*EC (dSm^{-1})*	*OC ($gmkg^{-1}$)*	*Available N ($kgha^{-1}$)*	*Available P_2O_5 ($kgha^{-1}$)*	*Available K_2O ($kgha^{-1}$)*
1	Noorpur	20	7.6	0.318	5.7	195.46	21.79	210.67
2	Adampur	15	7.9	0.660	5.3	219.96	32.85	205.50
3	Bhoor	20	8.0	0.582	6.6	255.90	28.16	190.39
4	Bhopur Bhaipur	15	8.5	0.365	5.4	208.88	41.83	207.30
5	Dhakia Jatt	15	7.8	0.218	6.5	250.36	45.39	226.60
6	Dharaknagla	20	7.0	0.785	5.8	218.62	36.21	158.20
7	Darapur	15	7.2	0.955	6.7	245.23	29.68	178.60
8	Fatanpur	15	6.6	0.155	5.7	225.00	44.16	205.30
9	Dheengapur	20	6.8	0.631	6.9	235.66	46.53	283.25
10	Fazalpur	12	7.5	0.251	6.5	248.88	38.33	265.60
11	Gamri	10	7.3	0.811	4.9	209.44	45.16	272.80
12	Isapur	15	8.2	0.456	4.8	196.88	42.28	214.72
13	Jatpura	20	8.3	0.254	3.9	156.96	41.81	212.71
14	Malpura	10	5.9	0.531	6.8	259.32	29.69	206.52
15	Manawala	15	6.7	0.765	5.4	226.53	28.23	160.15
16	Lalabala	20	9.2	0.120	5.9	228.78	35.42	189.20
17	Ratupura	12	9.0	0.989	6.8	246.56	38.43	201.32
18	Shivnagar	15	8.4	0.553	5.6	210.52	42.81	198.05
19	Zafrabad	10	6.1	0.456	6.7	246.48	46.21	220.20
20	Rosanpura	20	5.8	0.622	5.3	216.59	38.23	252.60
21	Padla	12	5.2	0.287	4.5	208.72	27.70	190.72
		Range	5.2- 9.2	0.120-0.989	3.9 – 6.9	156.96 – 259.32	21.79 – 46.53	158.20 – 283.25
		Mean	7.3	0.512	5.79	224.32	37.18	211.92

Available Phosphorous Content

The available phosphorous content in rice- wheat varied from 21.79 to 56.53 P_2O_5 kg ha^{-1} with a mean value of 37.18 P_2O_5 kg ha^{-1}. On the basis of the limit suggested by Muhr *et al.* (1963), 92 per cent samples were medium (20 to 50 P_2O_5 kg ha^{-1}) and 10 per cent were high (>50 P_2O_5 kg ha^{-1}) in available phosphorus. A significant positive correlation (r = 0.683) was observed between organic carbon and available phosphorous. A significant positive correlation (r = 0.684) was observed between organic carbon and available phosphorus. This relationship might be due to the presence of more than 50 per cent of phosphorous in organic form and after decomposition of organic matter as humus is formed which forms complex with Al and Fe and that is positive cover for P fixation with AL and Fe thus reduce phosphorus fixation (Tisdale *et al.*, 1997).

Available Potassium Content

Status of available potassium in the soil in rice- wheat ranged from 158.20 to 283.25 K_2O Kg ha^{-1} with an average value of 211.92 K_2O kg ha^{-1}. According to limit suggested by Mahr *et al.* (1963), all samples were medium (125 to 300 K_2O kg ha^{-1}) in potassium content. A significant positive correlation (r =0.615) was observed between organic carbon and available potassium. This might be due to creation of favorable soil environment with presence of high organic matter. Similar result was also reported by Paliwal (1996)

Micronutrients

Copper

The DTPA extractable Cu in the surface soil in rice- wheat of 21 grame panchayat was found to sufficient and varied from 0.258 to 1.708 mg kg^{-1} soil in surface (0-15cm) with a mean value of 1.095. All the observed values were well above the critical limit of 0.20 mg kg^{-1} as proposed by Lindsay and Norvell (1998).

Table 21.2. DTPA-Extractable Micronutrients (Cu, Fe, Mn and Zn) Status of Soil under Rice-Wheat Farming System

Sl. No.	Name of Village	No of Samples Collected	Cumg/kg	Femg/kg	Mnmg/kg	Znmg/kg
1	Noorpur	20	0.845	16.852	3.485	1.481
2	Adampur	15	0.854	11.258	4.621	0.435
3	Bhoor	20	0.258	8.296	3.426	0.810
4	BhopurBhaipur	15	0.881	13.652	5.514	1.708
5	Dhakia Jatt	15	0.923	6.145	4.156	0.835
6	Dharaknagla	20	0.869	6.253	4.152	0.823
7	Darapur	15	0.748	8.472	5.593	0.543
8	Fatanpur	15	1.028	3.214	1.916	0.781
9	Dheengapur	20	1.312	4.785	2.891	0.798
10	Fazalpur	12	2.415	5.596	2.245	1.839
11	Gamri	10	0.793	4.985	2.561	0.575
12	Isapur	15	2.041	10.581	8.351	1.685
13	Jatpura	20	0.916	5.596	2,856	1.069
14	Malpura	10	1.178	4.543	4.561	0.878
15	Manawala	15	0.620	6.391	5.186	0.595
16	Lalabala	20	1.541	4.045	5.170	0.583
17	Ratupura	12	0.623	6.379	2.267	0.861
18	Shivnagar	15	1.216	11.273	5.061	0.425
19	Zafrabad	10	1.368	3.738	1.701	0.768
20	Rosanpura	20	0.869	7.560	2.746	0.756
21	Padla	12	1.708	4.746	3.216	0.665
		Range	0.258- 1.708	3.214 - 16.852	1.701 – 8.351	0.425 - 1.708
		Mean	1.095	7.350	3.889	0.900

Iron

The DTPA -extractable Fe in the surface soil in rice- wheat of 17 grame panchayatwas was to be sufficient and varied from 3.214 to 16.852 mg kg^{-1} with a mean value of 7.350 mg kg^{-1}. According to critical limit of 4.5 mg kg^{-1} soil as suggested by Lindsay and Norvell (1978).

Mn

The DTPA- extractable Mn in surface soil varied from 1.701 to 8.351 mg kg^{-1} soil of 326 locations under rice- wheat farming system and is sufficient to high since are well above according to critical limit of 1.0 mg kg^{-1} as proposed by Lindsay and Norvell (1978).

Zn

The available Zn in surface (0-15 cm) in rice- wheat ranged from 0.425 to 1.708 mg kg^{-1} soil. According to critical limit 0.6 mg kg^{-1} as proposed by Lindsay and Norvell (1978) all the surface soils with exception of Adampur, Gamri and Manwala, Lalabala and Shivnagar were sufficient in Available Zn content.

Conclusion

The study of soil samples reveals that the soil of Thakudwara Tehsil of Moradabad district were did not followed a particular pattern with different grame panchayat which may be due to variation in management practices and yield potential. Nutrient status regarding to the available macro and micro nutrient in surface soil indicate that soils are low in available N and medium in available P and K and in general marginal in available Cu,Fe, Mn and Zn. Normal to slightly alkaline in reaction, low to medium in organic carbon content (Kumar et.al. 2013).

REFERENCES

Aggarwal, R.K., Kumar, P. and Sharma, B.K. (1990). Distribution of nitrogen in some Aridisols. *Journal of the Indian Society of Soil Science* 38, 430-433.

Brady, N.C. (1985). The nature and properties of soil, 8th edition Macmillan publishing Co. Inc., New York.

Jacksion M L (1973). Soil Chemical Analysis. Prentice Hall of India Pvt. Ltd.,New Delhi.

Kumar Ravindra, Rathi, A.S, Kant,L, Tripathi, N.C. and Pramod Kumar (2013). Soil fertility status of soil of Rampur District of Uttar Pradesh. *Journal of Krishi Vigyan Kendra*, 2(1): 55-58.

Kumar, P. Kumar, A. Dhyani, B.P., Kumar, P., Shahi, U.P., Singh,S.P., Kumar, Ravindra, Kumar, Yogesh (2013). Soil fertility status in some soils of Muzaffarnagar district of Uttar Pradesh,India, along with Ganga Canal command area. *African Journal of Agricultural Research*,Vol8(14),pp+1209-1217,18 April, 2013.

Lindsay, W.L. and Norvell, W.A. (1978). Development of DTPA soil test for zinc, iron, manganese and coper. *Soil Science Society of America Journal*, 42: 421-428.

Muhr, G.R., Datta, N.P., Sharma, S.N. Derer, F., Lecy, V.K. and Donahue, R.R. (1963). Soil testing in Indian, USAID mission to India.

Mishra,B. (2005). Soil quality and agricultural sustainability. Training held at Dept. of Soil Science, GBPUA and T, Pantnagar, from 01-12-2005 to 21-12-2005.

Olsen, S.R., Cole, C.V.,Watanabe, F.S. and Deen, L.A.(1954). Estimation of available phosphorus in soils by extraction with sodium bicarbonate. USDA CIRC. 939. Unied State Dept. of Ag. Wasignton.D.C.

Paliwal, M.L. (1996). Studies on major and micronutrient status of soils of Panchayat Samiti Bhinder, Udaipur. M.sc. (Ag) thesis, Rajasthan Agricultural University, Bikaner.

Ratan, R.K. and Sharma, P.D. (2004). Main micronutrient available and their method of use. Proceeding of IFA International Symposium on micronutrients. 110

Subbiah, B.V. and Asija, G.L. (1956). A rapid procedure for the determination of available nitrogen in soil. *Current Sci.* 25, 259-260.

Tisdale, S.L., Nelson, W.L., Beaton, J.D. and Havlin, J.L. (1997). *Soils fertility and fertilizer* 5th edition, Macmillan publishing Co. New Delhi 144-180, 198-201.

Verma, L.P., Tripathi, B.R. and Sharma. D.P. (1980). Organic carbon as an index to assess the nitrogen status of the soil. *Journal of the Indian Society of Soil Science* 28, 138-140.

Walkley, A.J. and Black, I.A. (1934). Estimation of soil organic carbon by the chromic acid titration method. *Soil Sci.* 37, 29-38.

Yadav, J.S.P. (2003). Managing soil health for sustainable productivity.Journal of the Indian society of Soil Science 51,448-465.

Chapter 22

Implementation of Integrated Pest Management of Vegetables in and Around NCR of Delhi through Electronic Media: A New Thought

Meenakshi Malik and Mukesh Sehgal*

ICAR-National Research Centre for Integrated Pest Management, New Delhi – 110 012

The Bulandshahr district administration urge the farmers to grow pesticide-free vegetables and grow multiple crops in their fields. Among the vegetables one of the common is cabbage. Cabbage being the fourth most widely grown vegetable crop of the country, the IPM experts planned to carry out a study in the Bulandshahr district of Uttar Pradesh. India, having the natural advantage of diverse agro-climatic conditions that enables the production of a wide range of horticultural crops and is bestowed with rich bio-diversity, also the second largest producer of vegetables accounts for about 15 per cent of the total world's production. The total area under vegetable cultivation is around 10.1 million hectares in 2019 which is about 3 per cent of the total area in the country. An electronic package and android based mobile application on the crop would be developed for the vegetable crop for the study area for management and awareness of IPM implementation for strengthening the linkage development between the lab to the field hence can be implemented as a model and can be developed for other crops.The farmers of the region face tremendous pain to arrive at the decision for their crop right from the selection and raising to the final disposal and managing the crop throughout the cropping season.They get the scattered information in piecemeal mode and the more on an unscientific basis.

This approach can help the community to get the information in a better way as lack of proper and timely information as an authentic input is really lacking and missing from the ground reality and therefore they are forced to adopt and use whatever is provided by the seller or the sources.

Keywords: *Farmers, Bulandshahr, Electronic package, Cabbage, Vegetables.*

Cabbage (*Brassica oleracea* var. *capitata*) being an essential and important crop is good source of protein,contains all essential amino acids, particularly sulphur containing amino acids which is an excellent source of minerals such as calcium, iron, magnesium, sodium, potassium and phosphorus. It has substantial amount of β carotene provitamin A, ascorbic acid, riboflavin, niacin and thiamine. The major varieties of cabbage are Golden acre, Pusa Drum Head, Pride of India, Copenhagen Market, PusaMukta, Pusa Synthetic, Midseason Market, September Early, Early Drum Head, Late Large Drum Head and K-1. It is commonly cultivated in cool moist climate. It is grown as a winter crop in plains. It is grown in varied types of soils ranging from sandy loam to clay and requires a pH ranging from 5.5 to 6.5 for increasing the production.

Cabbage is the fourth most widely grown vegetable crop of the country and India is the leading country in the production of the vegetable crop.West Bengal, Odisha, Bihar, Assam, Karnataka, Haryana, U.P., Maharashtra and Gujarat are the major growers and a large variety of vegetables belonging to different groups are being cultivated in the country which includes solanaceous, cucurbitaceous, leguminous, Cole crops, root crops and leafy vegetables.

Being having the natural advantage of diverse agro-climatic conditions, Uttar Pradesh enables production of wide range of horticultural crops and is bestowed with rich bio-diversity and varied agro-climatic conditions which is ideal for growing a large variety of horticultural crops. The sector, which includes fruits, vegetables, floriculture, spices, medicinal and aromatic plants have gained significance in terms of enhanced income per unit area, leading to socio-economic improvement of the people of the state, whereas the vegetables including root and tuber crops occupy an important place in diversification of agriculture and have played a pivotal role in food and nutritional security of ever growing population of our country. The area production and productivity of horticulture crop has considerably increased as the state and central Govt. have focused attention towards these crops. More income per unit area and employment generation in short span of time have attracted the enterprising farmers of the state, which resulted in diversification towards horticulture crops. These crops have proved to be the boon to the small and marginal farmers of the state who accounts for more than 90 per cent holding of the state. This is more important because nearly 70 per cent of the population is dependent on agriculture. India is the second largest producer of vegetables in the world (ranks next to China) and accounts for about 15 per cent of the world's production of vegetables. The significant achievements have been obtained in terms of production as the current production level is over 90 MT and the total area under vegetable cultivation is around 6.2 million hectares which is about 3 per cent of the total area under cultivation in the country. Potato, tomato, onion, cabbage and cauliflower account for around 60 per cent of the total vegetable production in the country.

Need of the e-package

Brassica is effected though the globe with aphids and aphids alone cause 9-96 per cent reduction in yield of the vegetable crop. In India, a total of 37 insect pests have been reported to feed on cabbage and among the insect pests. For the management of the notorious pests, farmers have no other choice but to spray insecticides which have their own adverse effects. Over 98 per cent of sprayed pesticides reach a destination other than their target species, because they are sprayed or spread within whole agricultural ecosystem. Continuous application of similar pesticides may increase pest resistance and out beak of secondary pests or resurgence. The monitoring of the pest population and their timely management is an important component of an area-wide pest control which overcome the problem. The abiotic factors including including temperature, relative humidity, rainfall greatly influence the population of insect pests. It is a well-known fact that the contribution of vegetables remains highest about 60 per cent in horticulture crop productions over the last five years and India is the world largest producer of many fruits and vegetables but there still exist huge gap between per capita demand and supply. Apart from the health improvements, the production of vegetables improves the economy of a country as these are very good source of income and employment. Due to inadequate facilities and information related to the crop, multiple times a large production is wasted.The government is encouraging the extension workers/farmers/researchers in the form of training's and research through which the commercial production can also be multiplied. Farmers of the concerned region face the problems in identification of Insect-pests and diseases related with the vegetable crops especially cabbage, being important commodity of the area. They face tremendous pain to arrive at the decision for their crop right from the selection and raising to the final disposal and managing the crop throughout the cropping season. Farming community get the scattered information in piece meal mode and the more on unscientific basis. Their approach is to get the information for their commodity for managing it in a better way, but lack of proper and timely information as an authentic input is really lacking and missing from the ground reality and therefore they are forced to adopt and use whatever is provided by the local sellers or the sources. The proper and scientific information is missing for their commodity and this incur harm rather improvement or profit to both their produce as well to the environment. Past strategy for development of the agriculture sector in India has focused primarily on raising agricultural output and improving food security. The net result has been a 45 per cent increase in per person food production which has made India not only food self-sufficient at aggregate level, but also a net food exporter country. There is need to search some direct measures to explicitly recognize the need to raise farmers income and promote farmers welfare. The farmers income remained low, which is evident from the incidence of poverty among farm households. The low and highly fluctuating farm income is causing detrimental effect on the interest in farming and farm investments, and is also forcing more and more cultivators, particularly younger age group to leave farming. This can cause serious adverse effect on the future of agriculture in the country. Time has come now to replace the existing and absolute technology *i.e.* time consuming, outdated with a new approach of Information era which could be more economical and beneficial. More specifically, the ICT based developed packages involves the conceptualization, design, development, evaluation and application of innovative ways to use communication technologies in the rural domain, with a primary focus on agriculture.

Tools and Techniques

A random sampling technique is adopted to select the villages and the farmers from the Sikandrabad block of Bulandshahr district of Uttar Pradesh *etc.* A total of 4 villages have been selected from the study area. The research work involves the development of electronic package and android based application for the selected vegetable crop for the western UP region especially Bulandshahr district of Uttar-Pradesh. The information on cabbage would be collected from the farmers of the study area through prescheduled questionnaires and interaction. The secondary data would be collected on various aspects as crop husbandry, economic aspects, pest scenario, market scenario, Input scenario, soil sampling, ETL of the crop, Insecticides and pesticides used for the vegetable crop from the published literature, reports, extension offices, SDO. The collected information would be compiledand the database would be developed for the same.The developed web application would be developed using PHP and SqL server and the android based mobile app would be available on the google play store.

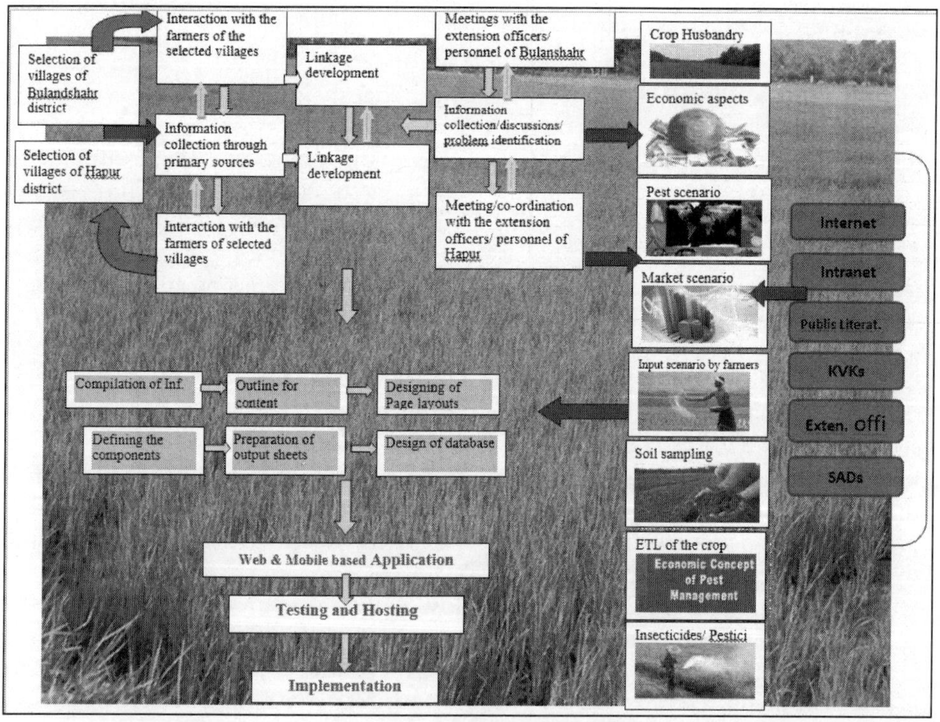

DFD for the Research Work

Data-flow diagram is a very important tool for ICT based applications that provide a graphical representation of the research work and aims to be accessible to the end users. The developed model enable the researcher and the team to work effectively during the analysis and specification of requirements. The technique of decomposition of high level data flow diagrams to a set of more detailed diagrams provides a complete view of system and the individual activities are described with more clarification for the better understanding of the user.

Conclusion

The e-package and the mobile technologies in the agriculture would provide much greater flexibility and capacity than paper-based systems, allowing for a more stream-lined integration of knowledge from different sources within a single-centred approach. It could be a surprise for most people that mobile technologies have reached the most traditional spheres of the Indian agriculture. As the technologies tend to simply multiple activities in various fields and also agriculture has become technology oriented. The agricultural productivity, agricultural-based problems such as food demand and build farms are more connected with the help of smart agriculture. The ICT technologies are emerging day by day and have become world's most used technologies of transferring data, voice and also various sort of services in the world. The information required by the farmers reaches in a effective, faster and cost-effective mode which helps to boost up with all the needs of the farmers.The online developed applications have given many benefits to the end-users starting from better land management judgements to quality yield. Farmers are using different types of apps to review the health of the crops during the crop cycle. Some of the mobile applications are developed to help the farmers in multiple ways like horticulture, crop management, weather forecast, agricultural field opportunities, expert suggestions, quality of soil, utilisation of fertilisers *etc.*The impact and sustainability of the efforts could be considerably enhanced through the use of more efficient tools and system for training and for integrating and delivering information to end users. The Bulandshahr district of Uttar Pradesh has been selected for implementation of the project as Cabbage is one of the most popular winter crop grown in this district and also the cultivation of vegetables in peri-urban or urban areas is recognized for its potential role in increasing food security, employment and income generation, poverty alleviation and environmental sustainability. There is an opportunity to build on existing knowledge by integrating information on the identification and management of insect pests, diseases, disorders and weeds within one flexible e-package using an Integrated Crop Management approach.

REFERENCES

Agarwal, P.K. *et al.* (1999). "Incidence of aphid, *Lipaphis erysimi* (Kalt.) on some cruciferous crop and chemical, control in cauliflower". *Indian journal of Applied Entomology*, 4: 19-25.

Babu, M. Surendra Prasad *et al.* (2010). "Implementation of Artificial Bee Colony (ABC) Algorithm on Garlic Expert Advisory System". *International Journal of Computer Science and Research*, 1(1).

Devraj, B. *et al.* (2011). "PulsExpert: An expert system for the diagnosis and control of diseases in pulse crops", *Expert Systems with Applications*, 38(9): 11463-11471.

El-Sayed *et al.* (2000). "Pest Control Expert System for Tomato (PCEST)". *Knowledge and Information Systems*, 2: 242-257.

Kumar, J. *et al.* (2016) "Infochemicals: An effective and environmentfriendly management of insect pests for sustainableagriculture". *International Journal of Agricultural Invention*, 1: 218-224.

Kumar, J. *et al.* (2017). "Differential electroantennogram response of male andfemale of Dysdercuscingulatus to okra plant volatiles". *Journal of Entomology and Zoology Studies*, 5: 197-201.

Kumar, J. *et al.* (2017). "Population dynamics of aphid, *Myzus persicae* (Sulzer) (Hemiptera: Aphididae) on different *Brassica* species." *Agricultural Science Digest*, 37: 64-67.

Prasad, M.S. *et al.* (2010). "A Web Based Tomato Crop Expert Information System Based on Artificial Intelligence and Machine Learning Algorithms". *International J. Computer Science and Information Technologies*, 1(1): 6-15.

Singh, Y.P. *et al.* (2012). "Integrated approach to managethe mustard aphid, *Lipaphis erysimi* (Kalt) (Homoptera: Aphididae) in oil seed Brassica crops-A review". *J. Aphidology*, 16: 77-88.

Yadav, V.K. *et al.* (2012). "Maize AGRIdaksh: A Farmer Friendly Device". *Indian Research Journal of Extension Education*", 12(3): 13-17.

Chapter 23

Sahiwal Cattle, The Pride of India: An Overview

***Devesh Singh*[1], *B.S. Khadda*[2] *and C.B. Singh*[3]**

[1]*International Institute of Veterinary Education and Research, Rohtak, Haryana*
[2]*ICAR-Krishi Vigyan Kendra- Panchmahal, Vejalpur, Godhra – 389 340, Gujarat*
[3]*G.B. Pant University of Agriculture and Technology, Pantnagar – 263 145, Uttarakhand*

Livestock is an important and integral part of agriculture sector in India. It plays an important role in the national economy and in the socio-economic development of the country. India mainly as an agriculture based nation where a large portion of population practice farming as a main profession, livestock gives a major contribution in sustaining rural economy of the country. India possesses the largest number of livestock population (20th Census, 2019) in the world as well as its milk production is the highest around the globe. Total cattle population in India is 192.49 million (35.94 per cent of total livestock population) and India contributed approximately 22 per cent of the total milk production of the world total production. As per the National Bureau of Animal Genetic Resources India have 50 well defined breeds of indigenous cattle. Among them Sahiwal is considered as the best milch dairy breed of the country and is well adapted to the tropical and sub tropical conditions of Indian sub continent. It is recognized by various names as Lola, Montgomery, Lambi-Bar and Multani. It reared by many farmers and dairy owners in India. The original breeding tract of this breed is in the central and southern area of Punjab particularly in Montgomery district in Pakistan. Small herd of Sahiwal cattle are found along the Indo-Pak borders of Ferozpur and Amritsar districts of Punjab and Sri Ganganagar district in Rajasthan. It is also found in some areas of UP and Chattisgarh. Sahiwal is known for low maintenance cost, high milk constituents, disease resistance and endurance to hot climatic conditions of tropics. Average lactation yield of Sahiwal cows is 2325 kilo grams. The lactation yield ranges from 1600 to 2750 Kg. However, selected herds may have higher productivity. Milk yield as high as 6000 lit has been recorded under organized farm conditions. The Age at first calving, first calving interval, service period, first lactation length, first lactation milk yield, 305 days first lactation milk yield and dry period of Sahiwal cattle were ranged 1101.07±8.53 days to 1390.00±3.90 days, 388.4±17.4 to 525.84±7.81 days, 119±3.31 days to 123.73±3.64, 205.98±5.32 days, 1941.59±58.76 kg, 1365.93±81.46 to 2030.19±19.96 and 75.3 days to 272.97±27.82days, respectively. Hence, this dairy breed has higher demand among the indigenous breeds. It is being widely used for the improvement of local stock through upgrading in some state of India. Sahiwal germ-plasm has been imported by more than 17 countries for improvement of their local stocks from India. It has been used for the production of different strains all over the world such as Jamaica hope, Australian milking Zebu, Australian Friesian Sahiwal (AFS) and Patina crosses. In India, Karan Swiss and Frieswal have been developed by crossing Sahiwal with Brown Swiss and Holstein Friesian, respectively. The Sahiwal cattle has an immense production potential with special reference to milk production however, their maximum potential has not been realized due to subsistence production system. The maximum production potential can be achieved by following the cattle improved husbandry practices, which will not only improve the productive and reproductive performance of Sahiwal cattle but also increase the socio-economic conditions of the dairy farmers.

Keywords: *Sahiwal, Milk production, Lactation period.*

Livestock is an important and integral part of agriculture sector in India. It plays an important role in the national economy and in the socio-economic development of the country. India mainly as an agriculture based nation where a large portion of population practice farming as a main profession, livestock gives a major contribution in sustaining rural economy of the country. India possesses the largest number of livestock population (20th Census, 2019) in the world as well as its milk production is the highest around the globe. Total cattle population in India is 192.49 million (35.94 per cent of total livestock population) and India contributed approximately 22 per cent of the total milk production of the world total production. As per the National Bureau of Animal Genetic Resources India have 50 well defined breeds of indigenous cattle. Among them Sahiwal is considered as the best milch dairy breed of the country and is well adapted to the tropical and sub tropical conditions of Indian sub continent. It is recognized by various names as Lola, Montgomery, Lambi-Bar, Multani and Teli. It reared by many farmers and dairy owners in India. Hence, this dairy breed has higher demand among the indigenous breeds. It is being widely used for the improvement of local stock through upgrading in some state of India. Sahiwal germ-plasm has been imported by more than 17 countries for improvement of their local stocks from India. It has been used for the production of different strains all over the world such as Jamaica hope, Australian milking Zebu, Australian Friesian Sahiwal (AFS) and Patina crosses. In India, Karan Swiss and Frieswal have been developed by crossing Sahiwal with Brown Swiss and Holstein Friesian, respectively.

Breeding Tract of Sahiwal Cattle

The breeding tract of this breed is mainly Ferozpur and Amritsar districts of Punjab and Ganganagar district of Rajasthan, however original breeding tract of this breed belong to central and southern area of Punjab particularly in Montgomery district in Pakistan. It is also found in some areas of UP and Chattisgarh. Good herds of pure Sahiwal cattle are available around Fazilka and Abohar towns of Ferozepur district in Punjab.

Characteristics of Sahiwal Cattle

The cows are brownish red in colour; shades may vary from a mahogany red brown to more greyish red. Extremities in bulls such as the head, legs and tail, are darker than rest of body colour. Occasionally there are white patches. Animals have well developed udder, good slope at rump and shorter horns. Dewlap is medium in size. Sahiwal is known for low maintenance cost, high milk constituents, it is also tick-resistant, heat-tolerant to hot climatic conditions of tropics and noted for its high resistance to parasites, both internal and external. The Sahiwal is the heaviest milker of all Zebu breeds and display a well-developed udder. Sahiwal demonstrate the ability to sire small, fast-growing calves and are noted for their hardiness under unfavorable climatic conditions.

Population Dynamics of Sahiwal Cattle in India

In India, contribution of indigenous cattle in the total cattle population is about 79 per cent. Of the Non-Descript category animals which is 73.83 per cent of the total Indigenous cattle. Out of the remaining 26.17 per cent, 4.2 per cent contribution is from Haryana breed followed by Gir and Sahiwal contributed over 3 per cent each of total indigenous cattle (Breed Survey, 2013 and Livestock Census 2019). Total Sahiwal population in India is 4,88,294.00 which is the third highest cattle population of India after Hariyana and Gir (**Breed Servey, 2013).** Sahiwal herds mainly maintained at University Dairy Farm Pantnagar, NDRI Karnal, Government Livestock Farm Hisar, Military Dairy Farm Meerut, Government cattle breeding farm Anjora, Durg and few other private and government farms *etc.*

Performance of Sahiwal Cattle

The reproductive and productive traits are most important traits of a dairy animal. Sahiwal cattle breed is a one of the best Zebu cattle breed that can particularly play same role as Holstein Friesian in tropical environment. The life time story of Sahiwal cattle represent an important genetic resources on which a total research has been carried out but it still out awaits genetic important to satisfy the national and international need for milk production. Some productive and reproductive traits of Sahiwal cattle are given below:

The Age at First Calving

The age at first calving determines the attainment of physical and sexual maturity. Economically, early first calving will ensure smaller investment and quicker return of capital. Genetically, it reduces the generation interval resulting in a better annual genetic gain from selection. A reduction in age of first calving can be achieved through better feeding, management, disease control and efficient heat detection and timely service programme. The average age at first calving of Sahiwal cattle was found to be 1281.89±15.57. The range value of AFC was 1219.00 ±19.77 to 1368.14 ±35.74 days.

Calving Interval

Calving interval is a most important economic trait of cattle. An ideal farm is run with the aim to reduce calving interval up to one year. The mean value of FCI trait was very much affected due to managemental practice and environment changes. The value of first calving interval was observed as 426.70±8.53 days, which are ranges from 379.59±18.71to 481.30±14.15 days. Calving interval for Sahiwal cows in was reported to be 388 to 498 days.

Service Period

Service period is the time period between date of calving and date of successful conception. The optimum service period helps the animal to recover from the stress of calving and also to get back the reproductive organs back to normal for cattle the optimum service period is 60-90 days. Efforts to reduce service period can be made by evaluating days open, efficiency of estrus detection and occurrence of conception. A problem with any of these will result in extended service period. The first service period of Sahiwal cattle was observed as 140.85±8.90 days (Devesh Singh, 2017). The average service period for Sahiwal cows was reported to be 73 (Bhatnagar et al., 1979) to 294 (Gaur and Raheja, 1996) days by various workers in India.

Lactation Length

The length of milk producing period after calving in lactation is known as lactation length. The optimum lactation length is 305 days. The milk production will breeding of dairy animals and farm records will be less, if this period is shortened. Indian breeds will have less lactation period, but in some breeds this period is more with very little milk production. The means value of first lactation length was found to be 290.42±5.57 days (Devesh Singh, 2017) which ranges from 280.51±7.18 to 307.11±7.8 days. The average lactation length in Sahiwal cattle was found to be ranged 247 (Khan *et al.*, 1999) to 348 (albott *et al.*, 1997) days.

Lactation Milk Yield

The least square mean of first lactation milk yield in the present study was estimated as 1927.50±36.68 kg which was ranges from 1430.19±60.85 to 2248.48±80.42 kg in Sahiwal cattle. The effect due to sire was found highly significant on the FLMY. It indicated that superior sire could be used effectively for the improvement of this trait. Average lactation yield of Sahiwal cows is 2325 kilo grams. The lactation yield ranges from 1600 to 2750 Kg. However, selected herds may have higher productivity. Milk yield as high as 6000 lit has been recorded under organized farm conditions.

Dry period: Dry period is the period from the date of drying (stop of milk production) to next calving. When the animal in pregnancy, before next calving. The animal should be given rest period to compensate for growth of foctus. A minimum of 2 months dry period should be allowed) If the dry period is not given or too low dry period, the animals suffer from stress and in next lactation, the milk production drops substantially and also it gives weak calves. On the other hand if the dry period given is too high, it may not have that much effect on increasing milk yield in the next lactation, but it decrease the production in the present lactation. The mean value of first dry period in Sahiwal cattle was estimated as 136.28±8.78 days and it ranges 83.62±19.25 to 194.58±14.57 days. The dry period in Sahiwal cattle was ranges 127 to 148 days.

Conclusion

The Sahiwal cattle has an immense production potential with special reference to milk production however, their maximum potential has not been realized due to subsistence production system. The maximum production potential can be achieved by following the cattle improved husbandry practices, which will not only improve the productive and reproductive performance of Sahiwal cattle but also increase the socio-economic conditions of the dairy farmers.

REFERENCES

BAHS (2019). Basic Animal Husbandry Statistics-2019. Ministry of Fisheries, Animal Husbandry and Dairying, Department of Animal Husbandry and Dairying, Krishi Bhawan, New Delhi.

Bhatnagar, D.S., Taneja, V.K., Basu, B.S. andMurthy, K.M.K. (1983). Genetic parameters for some economic traits in Sahiwal cattle. *Indian J. Dairy Sci.*, 36: 402-406.

Breed Survey (2013). Estimated livestock population breed wise survey 2013. Government of India, Ministry of agriculture and farmers welfare, Department of animal husbandry, dairying and fisheries, (Animal husbandry statistics division) Krishi Bhawan, New Delhi.

Gaur, G.K. and Raheja, K.L. (1996). Best line arun biased estimates of different environmental effects for production and reproduction traits in Sahiwal. *Indian J. Anim. Sci.*, 66: 603-606.

Livestock Census (2019). *The 20th Livestock Census, 2019.* Ministry of Fisheries, Animal Husbandry and Dairying, Government of India, Krishi Bhawan, New Delhi.

Khan, M.S. and Iqbal, J. (1999). Lactation length adjustment for milk yield records: A review. *Int. J. Agri. Biol.*, 1(3): 199-204.

Manoj, M., Gandhi, R.S., Raja, T.V., Singh, A. and Sachdeva, G.K. (2012). Effect of non-genetic factors on the first lactation production and reproduction performance in Sahiwal cattle. *Indian journal of Dairy Science*, 65(3): 264-265.

Raja, K.N. (2004). Genetic and phenotypic trends for production and reproduction performance of Sahiwal cattle. M.V.Sc. Thesis, NDRI Deemed University Karnal, India.

Singh, Devesh (2017). M.V.Sc. Thesis submitted to to the G. B. Pant University of Agriculture and Technology Pantnagar-263145 (U.S.Nagar) Uttarakhand (India).

Singh, V.K., Singh, C.V., Kumar, D. and Kumar, A. (2005). Genetic evaluation of some economic traits in Sahiwal and its crosses. *Indian J. Dairy Sci.*, 58 (3): 206-210.

Singh, J. and Singh, C.V. (2016). Genetic and phenotypic parameters of first lactation and life time traits in Sahiwal cows. *J. Vet. Sci. Tech.*, 7 (4): 345.

Talbott, C.W., Chauhdry, M.Z., McDowell, R.E. and McDaniel, B.T. (1997). Potential to increase milk yield in tropical countries with indigenous dairy cattle: The Sahiwal Model. *J. Anim. Plant Sci.*, 7(1-2): 1-10.

Index

E

F

G

H

I

K

L

M

N

O

P

R

S

T

V

W